GENE TRANSCRIPTION: DNA BINDING PROTEINS

ESSENTIAL TECHNIQUES SERIES

Series Editor

D. Rickwood

Department of Biological and Chemical Sciences, University of Essex, Wivenhoe Park, Colchester, UK

Published titles

Antibody Applications
Gel Electrophoresis: Nucleic Acids
DNA Isolation and Sequencing
Gel Electrophoresis: Proteins
PCR
Gene Transcription: DNA Binding Proteins
Gene Transcription: RNA Analysis

Forthcoming titles

Human Chromosome Preparation
Cell Biology
Antibody Production
Vectors: Cloning Applications
Cell Culture
Vectors: Expression Systems
Nucleic Acid Hybridization

GENE TRANSCRIPTION: DNA BINDING PROTEINS

ESSENTIAL TECHNIQUES

Edited by

K. Docherty

Department of Molecular and Cell Biology, University of Aberdeen, Aberdeen, UK

JOHN WILEY & SONS

Chichester • New York • Brisbane • Toronto • Singapore

Published in association with BIOS Scientific Publishers Limited

British Library Cataloguing in Publication Data
A catalogue record for this book is available from the British
Library.

Library of Congress Cataloging in Publication Data
A catalogue record for this book is available from
the Library of Congress.

ISBN 0 471 97016 6

Typeset by Footnote Graphics Ltd, Warminster, UK
Printed and bound in UK by Biddles Ltd, Guildford, UK

CONTENTS

Contents

CONTRIBUTORS

Admon A. Technion, Israel Institute of Technology, Technion City, Haifa 32000, Israel

Allan J. Department of Biochemistry, Edinburgh University, George Square, Edinburgh EH8 9XD, UK

Belyaeva T. School of Biochemistry, University of Birmingham, PO Box 363, Birmingham B15 2TT, UK

Busby S. School of Biochemistry, University of Birmingham, PO Box 363, Birmingham B15 2TT, UK

Clark A.R. Imperial Cancer Research Funds, Lincoln's Inn Field, London, UK

Docherty K. Department of Molecular and Cell Biology, University of Aberdeen, Institute of Medical Sciences, Aberdeen AB25 2ZD, UK

Goodbourn S. Division of Biochemistry, St George's Hospital Medical School, London SW17 0RE, UK

Hammond-Kosack M.C.U. Department of Molecular Genetics, Cambridge Laboratory, John Innes Centre, Colney Lane, Norwich NR4 7UJ, UK

Macfarlane W.M. Department of Molecular and Cell Biology, University of Aberdeen, Institute of Medical Sciences, Aberdeen AB25 2ZD, UK

Moore P.A. Department of Molecular and Cell Biology, University of Aberdeen, Institute of Medical Sciences, Aberdeen AB25 2ZD, UK

Philippe J. Diabetes Unit, University Hospital, 1211 Geneva 14, Switzerland

Read M. Department of Molecular and Cell Biology, University of Aberdeen, Institute of Medical Sciences, Aberdeen AB25 2ZD, UK

Savery N. School of Biochemistry, University of Birmingham, PO Box 363, Birmingham B15 2TT, UK

Walker M.D. Biochemistry Department, Weizman Institute of Science, Rehovot 76100, Israel

ABBREVIATIONS

APS	ammonium persulfate		LB	L-broth
AT	amino triazole		LCR	locus control region
			LMPCR	ligation mediated PCR
bp	base pair			
BSA	bovine serum albumin		ORF	open reading frame
CBP	CREB-binding protein		PBS	phosphate buffered saline
CREB	cyclic AMP response element binding factor		PCR	polymerase chain reaction
			PMSF	phenylmethylsulfonyl fluoride
DMS	dimethyl sulfate		PVDF	polyvinylidene difluoride
dNTP	deoxyribonucleotide			
DTT	diothiothreitol		r TAT	rat tyrosine amino transferase
EDTA	ethylenediamine tetra-acetic acid		SDS	sodium dodecyl sulfate
EGTA	ethylenebis tetra-acetic acid		SHC	SSC hybridization buffer
EMSA	electrophoretic mobility shift assay		SSC	standardized saline citrate
EST	expressed sequence tag			
			TBB	Tris-binding buffer
GHCl	guanidine hydrochloride		TE	10 mM Tris-HCl pH 8.0, 1 mM EDTA
GST	glutathione S transferase		TEMED	N,N,N',N'-tetramethylenediamine
			THB	TMAC hybridization buffer
HBB	hepes binding buffer		TMAC	tetramethylammonium chloride
HSV tk	thymidine kinase of Herpes simplex virus		TWB	TMAC wash buffer
IPTG	isopropyl-β-D-thiogalactopyranoside			

PREFACE

The past 10 years or so have seen major advances in our understanding of the mechanisms involved in the regulation of gene transcription. The basic mechanisms are similar in both bacteria and eukaryotic organisms (i.e. a transcriptional complex comprising RNA polymerase and a number of additional protein factors associate with the gene at a sequence, the promoter, located close to the mRNA start site). The activity of this transcriptional complex is regulated both positively and negatively by transcription factors that bind to multiple *cis*-acting sequences located either close to or at a distance from the promoter. The net activity of these transcription factors is dependent on extracellular signals, usually either nutrient metabolites or hormones. In multicellular organisms the activity of transcription factors can determine at what stage, and where, in the developing embryo a particular gene is expressed, while in differentiated tissue they act to restrict gene expression to particular cell types.

This book describes techniques used in the study of transcription factors and DNA sequences involved in regulating gene transcription. Chapter 1 describes the various polymerase complex at the promoter either directly or through an adapter protein. Much of this information has come from advances in techniques enabling us to clone and sequence transcription factors (Chapter 3). With the rapid progress being made in sequencing the entire genome of a number of organisms we can expect that database searching by homology (Chapter 4) will reveal novel transcription factors for which physiological functions and target genes will be required. Understanding the role of these transcription factors in health and disease will be a major challenge of the next decade.

Methods used to map *cis*-acting sequences involved in regulating gene transcription are described in Chapter 5. These include the now classical Bal 31 deletion mutagenesis, along with more sophisticated techniques for introducing sequence specific deletions and targeting mutagenesis to specific base pairs. An outcome of studies employing these methods has been the observation that multiple sequences are involved, and that DNA-bound transcription factors interact with other factors both on and off the DNA. The yeast two-hybrid system (Chapter 6) is a techniques used to analyze protein–DNA interactions. These include the electrophoretic mobility shift assay which, with its relative simplicity combined with exquisite sensitivity and potential specificity, is used in most laboratories working in this field. Other more sophisticated footprinting techniques, whereby the exact nucleotides involved in interacting with proteins are mapped, are also described. The problem, however, has been in relating these *in vitro* findings to the situation in living cells. Specifically, how can one map proteins binding to a single gene species in a mixture of up to 10^5 different genes. This problem has now largely been overcome with the introduction of ligation mediated polymerase chain reaction (LMPCR) based methods (Chapter 2). Thus it is now possible to study the effect of hormone/metabolic signals on, or developmental aspects of, transcription factors binding to specific DNA sequences *in vivo*.

Transcription factors are for the most part composed of discrete domains involved in binding to DNA (e.g. the homeodomain), interacting with other transcription factors to form homodimers or heterodimers (e.g. the leucine zipper), or activation domains that interact with the RNA powerful method for studying in detail protein–protein interactions, and has been used to clone proteins that bind to known transcription factors. When combined with sensitive assays (Chapter 1) and cloning techniques (Chapter 2) based on DNA–protein interactions, the yeast two-hybrid system provides a powerful armoury for characterizing in detail all factors involved in gene transcription. Fine regulation of their activity in many cases involves phosphorylation/dephosphorylation reactions. Chapter 7 describes a number of techniques involved in characterizing the phosphorylation status of transcription factors, mapping phosphorylated amino acids, and identifying kinases. Finally, in Chapter 8 a method used to map hypersensitive sites in DNA is described. Such hypersensitive sites are present in all genes that are actively transcribed.

The other major techniques employed in studying transcriptional regulation such as transcript analyses, transfection of DNA into cells, use of reporter constructs and others, are described in the companion volume to this book, *Gene Transcription – RNA Analysis: Essential Techniques*.

Kevin Docherty

SAFETY

Attention to safety aspects is an integral part of all laboratory procedures and national legislations impose legal requirements on those persons planning or carrying out such procedures. While the authors, editor and publisher believe that the recipes and practical procedures, as set forth in this book, are in accord with current recommendations and practice at the time of publication, they accept no legal responsibility for any errors or omissions, and make no warranty, expressed or implied, with respect to material contained herein. It remains the responsibility of the reader to ensure that the procedures which are followed are carried out in a safe manner and that all necessary safety instructions and national regulations are implemented.

In view of ongoing research, equipment modifications and changes in governmental regulations, the reader is urged to review and evaluate the information provided by the manufacturer, for each reagent, piece of equipment or device, for any changes in the instructions or usage and for added warnings and precautions.

These are freely accessible using:
http://joule.pcl.ox.ac.uk/MSDS/.

Other safety information on the Internet can be accessed on:
gopher://atlas.chem.utah.edu/11/MSDS
gopher://ginfo.cs.fit.edu:70/lm/safety/msds
http://physchem.ox.ac.uk/MSDS
http://www.fisher1.com/Fischer/Alphabetical Index.html
http://www.pp.orst.edu

You are actively encouraged to check these data sheets to confirm our assignments and for more detailed information on individual hazards; however the author, editor and publisher can accept no responsibility for any material contained in these data sheets. Furthermore, you must always follow the precautions outlined on labels and data sheets provided by individual manufacturers.

Radiation. The use of radioisotopes is subject to legislation and requires permission in most countries. Furthermore, national guidelines for their use and disposal must be rigorously adhered to. The procedures in protocols that

All procedures mentioned within this book must be carried out under conditions of good laboratory practice in accordance with local and national guidelines. Some procedures involve specific hazards, including but not limited to hazards in the following categories:

Chemical. A number of the reagents are known to be carcinogenic, mutagenic, toxic, inflammable, highly reactive or otherwise hazardous. Substances known to be hazardous have been marked with the symbol ⚠ in the list of reagents (but not subsequently) for each protocol, or if they appear as alternatives to the main protocol, the *first time* they appear in the notes. The reader should consult the safety notes on these pages before embarking on any of the procedures covered. This is in no way meant to imply that undesignated chemicals are nonhazardous, and all laboratory chemicals should be handled with extreme caution. Information is not available on the possible hazards of many compounds. The criteria we have generally used for denoting a substance with ⚠ is based upon a hazard level of 2 or more (on a scale 0–4) in any of the categories in the Baker Saf-T-Data™ system used in the material safety data sheets (MSDS) held at the University of Oxford, UK.

use radioisotopes must only be carried out by individuals who have received training in the use of such material using the appropriate facilities, protection and personal monitoring procedures.

Biological. Antibodies, sera and cells (particularly, but not exclusively, those of human and nonhuman primate origin) pose a significant biological hazard. All such materials, whatever their origin, may harbor human pathogens and should be handled as potentially infectious material in accordance with local guidelines. Any recombinant DNA work associated with protocols is likely to require permission from the relevant regulatory body and you must consult your local safety officer before embarking upon this work.

Electrical. Many of the procedures in this book use electrical equipment. Electrophoresis techniques may present particular hazards of this nature.

Lasers. Flow cytometers and certain other types of laboratory equipment contain lasers. Users should ensure they are fully aware of the potential hazards of using such equipment.

I PROTEIN–DNA INTERACTIONS.   N. Savery, T. Belyaeva and S. Busby

Introduction

Interactions between proteins and DNA play a key role in the regulation of gene expression. Sequence-specific contacts between proteins and promoter DNA determine the site of transcription initiation with precision, and many of the transcription factors responsible for activating or repressing gene expression bind to DNA in a sequence-specific manner. The ability to determine how and where such transcription factors bind, and how RNA polymerase interacts with base sequences at different promoters is essential if the intricate workings of the transcriptional machinery are to be understood.

Electrophoretic mobility shift assays (EMSAs) (see *Protocol 1*)

Many techniques are now available to study protein–DNA interactions *in vitro*; one of the simplest is the electrophoretic mobility shift assay (EMSA) [1]. The protein (or cell extract) of interest is mixed with radioactively labeled DNA and incubated to allow a complex to form. The mixture is then electrophoresed under nondenaturing conditions. Many protein–DNA complexes are surprisingly stable during electrophoresis and such complexes have a different (usually lower) electrophoretic mobility from free DNA. Electrophoresis thus separates the protein–DNA complexes from unbound DNA, and bands due to bound and unbound DNA can be detected by autoradiography. This procedure in its most basic form can be used to ascertain whether a protein can bind at a particular DNA site, or whether a cell extract contains a particular DNA-binding protein, but with a little modification it can be used to measure kinetic parameters and to assess the sequence specificity of the protein–DNA interaction being studied [2,3].

Interference assays (see *Protocol 2*)

More specific information about the way in which a protein interacts with DNA can be obtained from 'interference' assays in which the labeled DNA fragment is treated with a modifying agent before being mixed with the protein [4]. The modification is performed in conditions such that, on average, each individual DNA molecule is altered only once. A set of DNA molecules carrying modifications at different positions is thus generated. Some of these modifications will occur at bases which are essential for interaction with the DNA-binding protein, and fragments carrying such critical modifications will remain unbound when the DNA is incubated with the protein. After nondenaturing gel electrophoresis (as used in EMSA) the unbound molecules are extracted from the gel, cleaved at the site of modification and analyzed by denaturing gel electrophoresis. Comparison of the ladder of cleavage products with a suitable sequence ladder reveals the position of the bases at which modification interfered with the binding of the protein, and which are likely to be in close proximity to the protein when it is bound.

Southwestern blotting (see *Protocol 3*)

Southwestern blotting is a procedure which identifies DNA-binding proteins within a cell extract or a partially purified mixture of proteins, and it is particularly useful for the detection of previously undiscovered DNA-binding proteins capable of binding to particular nucleotide sequences [5,6]. The protein mixture is separated using either native or denaturing gel electrophoresis and the proteins are then electrophoretically transferred on to nitrocellulose filters (after a renaturing step, if denaturing gels were used). The nitrocellulose filters are probed with a highly radioactive DNA fragment carrying the nucleotide sequence of interest, and any proteins that bind that sequence are detected by autoradiography.

DNA footprinting (see *Protocols 4–6*)

A widely used set of techniques is based on 'footprinting', in which the location of a bound protein is detected by the protection it affords the DNA against chemical or enzymic attack. When an end-labeled DNA fragment is attacked by an agent which cleaves the DNA (preferably in conditions in which each molecule is cut only once) a set of radioactive DNA fragments of different lengths is generated. These will have one common (labeled) end, and the number and length of fragments produced will depend on the sequence specificity of the agent used for cleavage. If a bound protein prevents cleavage at one or more sites, a subset of the fragments will not be produced. When paired samples prepared from incubations in the presence and absence of DNA-binding protein are compared by denaturing gel electrophoresis, a number of bands will be missing. The resulting gap in the DNA ladder is termed the footprint of the protein. A number of different reagents are available for footprinting studies; in this chapter DNase I, the hydroxyl radical, and permanganate ions are considered.

DNase I is an enzyme which binds in the minor groove of the DNA duplex and cuts the phosphodiester backbone of both strands independently [7]. The enzyme shows some sequence specificity and a DNA ladder produced by partial DNase I digestion is uneven, with some regions being inefficiently cut. It is a relatively large molecule (approximately 40 Å in diameter) and steric hindrances prevent it from attacking the DNA close to another bound protein. Thus the extent of protection is often greater than the area actually bound by the protein being studied, which means that most DNA-binding proteins give clear footprints in DNase I protection assays. The technique is not, however, particularly informative for very high resolution mapping of protein–DNA interactions, particularly within multi-protein complexes such as the transcription complex. In contrast, the hydroxyl radical is much smaller than DNase I, is readily diffusible, and is able to access the DNA even within multi-protein complexes [8]. Hydroxyl radicals can be generated *in vitro* by the action of hydrogen peroxide on $[Fe(II)(EDTA)]^{2-}$, and can attack either single- or double-stranded DNA in a sequence independent manner. The primary target of hydroxyl radical attack is deoxyribose (hence its lack of sequence specificity) and cleavage of the sugar phosphate backbone is brought about by secondary reactions. A DNA ladder generated by hydroxyl radical cleavage thus contains a band corresponding to every position in the sequence and can give a much clearer picture of the manner in which a particular protein binds, and the way in which its binding can be altered by changes in DNA sequence or the addition of further protein factors. Effects of protein binding on the pattern of hydroxyl attack can, however, be quite subtle, often requiring computer analysis of the data. This is because the degree of protection afforded by any protein against hydroxyl radical attack is less than that which it can offer against a larger molecule such as DNase I.

Other footprinting reagents are specific for particular DNA structures; for example permanganate specifically recognizes single-stranded DNA and can be used as a probe for localized unwinding of the double helix, primarily in studies of open complex formation at promoter regions [9]. As with other footprinting techniques, permanganate is added to preformed DNA–protein complexes. It is a strong oxidant which reacts with bases; all bases can be modified to some extent, thymine is the most reactive. Permanganate modifies DNA but does not cleave it. Modifications are detectable because DNA polymerase is unable to copy past a permanganate-modified base, so primer extension from a radiolabeled primer followed by denaturing gel electrophoresis generates a ladder of DNA fragments with a band corresponding to each unpaired 'T' residue in the sequence (alternatively piperidine cleavage of an end-labeled permanganate-modified DNA fragment can be used to generate the ladder). This method of detection is very sensitive; localized unwinding of only a small percentage of the DNA can be readily detected as only modified DNA will lead to the production of truncated primer extension products. The primer extension methodology allows experiments on unlabeled or circular DNA, such as intact plasmids or chromosomal DNA, and therefore allows the effect of supercoiling on promoter opening to be examined. Additionally, it is very specific, the choice of primer determines the region of DNA which is examined and other DNA fragments present in the reaction mixture will not interfere.

Protocols provided

1. *Electrophoretic mobility shift assay (EMSA)*
2. *Methylation interference analysis*
3. *Southwestern analysis*
4. *DNase I footprinting*
5. *Hydroxyl radical footprinting*
6. *Potassium permanganate footprinting*
7. *Preparation of 'G' ladder by DMS treatment and piperidine cleavage*

References

1. Garner, M. and Revzin, A. (1981) *Nucleic Acids Res.* **9**:6505.
2. Fried, M. (1989) *Electrophoresis*, **10**:366.
3. Revzin, A. (1987) *Biotechniques*, **7**:346.
4. Siebenlist, U. and Gilbert, W. (1980) *Proc. Natl Acad. Sci. USA*, **77**:122.
5. Knepel, W., Jepeal, L. and Habener, J.F. (1990) *J. Biol. Chem.* **265**: 8725.
6. Tully, D.B. and Cidlowski, J.A. (1993) *Methods Enzymol.* **218**: 535.

7. Galas, D.J. and Schmitz, A. (1978) *Nucleic Acids Res.* **5**: 3157.
8. Tullius, T.D. and Dombroski, B.A. (1986) *Proc. Natl Acad. Sci. USA*, **83**:5469.
9. Sasse-Dwight, S. and Gralla, J.D. (1989) *J. Biol. Chem.* **264**:8074.

Reagents

40% (w/v) Acrylamide stock solution [prepare a 40% (w/v) solution of acrylamide: N,N'-methylene bisacrylamide (29:1, w/w) using Ultrapure grade chemicals and store at 4°C in the dark]. Caution: acrylamide is a neurotoxin, wear gloves ⚠

Adenosine 5'-[γ-^{32}P]triphosphate, triethylammonium salt (3000 Ci/mmol, 111 TBq/mmol) (Amersham)

20% (w/v) Ammonium persulfate (APS): usually made fresh as required but can be stored at 4°C for at least a month

Ammonium sulfate

5× Binding buffer [125 mM Tris-HCl pH 7.5, 250 mM KCl, 5 mM DTT, 25% (v/v) glycerol] store at 4°C

Buffer A [10 mM Hepes pH 7.9, 10 mM KCl, 1.5 mM MgCl$_2$, 0.5 mM dithiothreitol (DTT)]

Buffer B [20 mM Hepes pH 7.9, 10% (v/v) glycerol, 0.42 M NaCl, 1.5 mM MgCl$_2$, 0.5 mM DTT, 0.2 mM EDTA, 1 mM PMSF ⚠] add DTT and PMSF ⚠ just before use

Buffer C [20 mM Hepes pH 7.9, 10% (v/v) glycerol, 0.5 mM DTT, 0.2 mM EDTA, 1 mM PMSF ⚠]

Buffer D (10 mM Hepes pH 7.9, 10 mM KCl, 1× protease inhibitor cocktail solution)

Buffer E [20 mM Hepes pH 7.9, 400 mM NaCl, 5% (v/v) glycerol,

Mannheim, resuspend in 2 ml distilled water for a 25× stock solution)

Sephadex G-50 equilibrated in TE pH 7.6

Synthetic oligonucleotides

T4 polynucleotide kinase (usually supplied with 10× kinase buffer)

10× TBE (107.8 g Tris base, 55 g boric acid, 7.44 g EDTA) make up to 1 liter with deionized water, and store at room temperature

TE buffer pH 7.6 (10 mM Tris-HCl pH 7.6, 1 mM EDTA)

TEMED (N,N,N',N'-tetramethylethylenediamine), store at 4°C

Equipment

β-counter

B pestle for Dounce homogenizer

25–50 ml Beakers

Bench-top centrifuge

50 ml Centrifuge tubes

SW50 Centrifuge rotor (or equivalent)

5 ml Centrifuge tubes for SW50 rotor

Dialysis tubing

Dounce homogenizer

Dri-block at 90°C

Ice bucket

1× protease inhibitor cocktail solution]

Distilled water

0.5 M EDTA (pH 8.0)

10× EMSA gel loading buffer [50% (v/v) glycerol, 0.25% (w/v) xylene cyanol, 0.25% (w/v) Bromophenol blue]

10% Nonidet NP-40 solution

20× Oligo annealing buffer (200 mM Tris-HCl pH 7.9, 40 mM $MgCl_2$, 1 M NaCl, 20 mM EDTA) store at 4°C

Phenylmethylsulfonylfluoride (PMSF) stock solution (make up a 100 mM stock of PMSF in propan-2-ol and store at −20°C). Caution: PMSF is very toxic by inhalation and in contact with skin ⚠

Phosphate buffered saline (PBS)

Poly(dI:dC)-poly(dI:dC) (4 mg/ml in 100 mM NaCl) store at −20°C

Protease inhibitor cocktail tablet (e.g. Complete™, Boehringer

Magnetic stirrer

Microcentrifuge

Microcentrifuge tubes

Perspex radiation shield

Plastic film (e.g. Saranwrap)

Plastic spatula (rubber policeman)

Shaking platform

Slab gel dryer

Ultracentrifuge (Beckman L8 or equivalent)

Vertical polyacrylamide gel apparatus (BioRad Protean II xi cell or equivalent)

Vortex

Water baths at 37°C and 70°C

3MM Whatman paper

X-ray film

Procedure

Preparation of nuclear protein extracts

Method A

1 Cool buffers A, B and C to 4°C. ①

2 Harvest $1-5 \times 10^9$ cells and transfer to a 50 ml centrifuge tube. Wash the cells twice with 5 vols of PBS by centrifugation at 400 g for 10 min.

Notes

Time required: 2 days

Preparation of nuclear extract by method A – 24 h, including overnight dialysis. Method B 2–3 hours

Radiolabeling oligonucleotide probe – 3 h.

Electrophoretic mobility shift assay – 6–8 h depending on electrophoresis run time.

Protocol 1. Electrophoretic mobility shift assay

3 Resuspend the cells in 5 packed cell vol. of cold buffer A and allow to stand for 10 min.

4 Centrifuge at 400 g for 10 min at 4°C, and resuspend the cell pellet in 2 packed cell vol. of buffer A.

5 Lyse the cells on ice using a Dounce homogenizer with the B pestle. ②

6 Centrifuge at 400 g for 20 min at 4°C to pellet the nuclei. Remove the supernatant, taking care not to disturb the pellet which may be loose. ③

7 Resuspend the nuclei in 3 ml ice-cold buffer B/10^9 cells and homogenize with 20 strokes of the B pestle in a Dounce homogenizer.

8 Transfer the homogenate into a small beaker (25–50 ml) and stir the homogenate gently on a magnetic stirrer for 30 min at 4°C. Centrifuge at 100 000 g for 20 min at 4°C (32 700 r.p.m. in a SW50 rotor).

9 Transfer the nuclear supernatant to a small beaker (50 ml). Place the beaker on a magnetic stirrer at 4°C and precipitate the proteins by the gradual addition of solid ammonium sulfate to a final concentration of 0.33 g/ml.

10 Harvest the nuclear precipitate by centrifugation at 25 000 g for 20 min at 4°C (16 400 r.p.m. in a SW50 rotor). Resuspend the pellet in 0.5–1.0 ml buffer C/10^9 cells.

11 Dialyse against 100 vols of buffer C at 4°C for 12–18 h.

12 Transfer the dialysate to a 1.5 ml microcentrifuge tube(s) and centrifuge in a microcentrifuge for 10 min at 4°C. Store the supernatant at −70°C in 0.1–0.25 ml aliquots.

① A cocktail of protease inhibitors may be added to buffers B and C (e.g. 1× protease inhibitor cocktail solution) in addition to PMSF.

② 50–100 strokes are usually sufficient.

③ The supernatant may be retained to prepare a cytoplasmic S100 extract.

④ The spun-column is prepared by filling disposable columns (QS-Q, Isolab) with the G-50 slurry. Place the column in a 15 ml centrifuge tube and centrifuge for 5 min at 1200 r.p.m. (200 g). Carefully apply the sample on top of the dried matrix, and centrifuge for 5 min at 1200 r.p.m.. Apply 200 μl of TE to the surface of the resin and re-centrifuge.

⑤ The radioactivity should be in the order of 0.5–1 × 10^6 d.p.m./μl.

⑥ Measure the concentration of nuclear proteins by the Bradford method using a BioRad protein assay kit.

⑦ Polynucleotides are used to sequester proteins that bind nonspecifically to the probe. Typically, a range of polynucleotide concentrations from 0–2 μg are used to determine optimum binding conditions. A range of different polynucleotides should also be used, for example poly (dA:dT)-poly (dA:dT) as in some cases polynucleotides can sequester specific binding proteins.

⑧ Count 1 μl of probe diluted in 100 μl distilled water. Then dilute the probe to give an activity of approximately 320 c.p.m./μl.

⑨ Tris- or Hepes-based buffers may be used for the gel

Method B

1 Remove the growth medium and wash the cell monolayer with 10 ml PBS. Remove the wash and discard.

2 Add 1 ml TBS and dislodge the cells from the plate by scraping with a plastic spatula (rubber policeman).

3 Transfer the cells to a 1.5 ml microcentrifuge tube and centrifuge for 15 sec.

4 Remove the supernatant and resuspend the cells in 400 μl ice-cold buffer D. Allow the cells to swell on ice for 15 min.

5 Add 25 μl 10% Nonidet NP-40 solution, vortex vigorously for 15 sec and centrifuge for 30 sec.

6 Discard the supernatant and resuspend pellet in 50 μl ice-cold buffer E.

7 Vigorously rock the resuspended pellet at 4°C for 1–2 h on a shaking platform.

8 Centrifuge for 5 min at 4°C and transfer the supernatant in 10 μl aliquots to fresh tubes.

9 Snap freeze extract in liquid nitrogen and store at −70°C.

Radiolabeling the oligonucleotide probe

1 Set up the following reaction mix in a 1.5 ml microcentrifuge tube:
- 10× kinase buffer   1 μl
- sense oligonucleotide   20 pmol

retardation assay. Important factors are: (i) salt concentration – initially it is advisable to carry out a KCl/MgCl$_2$ titration; (ii) co-factor requirements - for example the detection of zinc finger protein complexes may require the omission of metal chelators from the buffer and supplementation with zinc.

(10) When using semi-purified or purified proteins Nonidet NP-40 (0.1%) can be included in the reaction buffer.

(11) Competition EMSAs may be used to investigate the specificity of a particular protein–DNA interaction. In this case the protein extract is pre-incubated with a 50–500-fold molar excess of unlabeled competitor oligonucleotide for 20 min before addition of the radiolabeled probe.

(12) Many protein–DNA interactions are unstable *in vitro* and are not easily detected using the basic EMSA method. Two ways of increasing the sensitivity of the EMSA method to detect such interactions include: (i) stabilizing the protein–DNA interaction by performing the EMSA in a cold room and/or (ii) reducing the overall run time of the EMSA by raising the voltage to 200–300 V (low ionic buffers and a peristaltic pump to recirculate the buffer are required when using higher voltages).

Protocol 1. Electrophoretic mobility shift assay

- [γ-^{32}P]ATP (3000 Ci/mmol,   5 μl (50 μCi, 1.85 MBq)
  111 TBq/mmol)
- T4 polynucleotide kinase (10 U/μl)   1 μl
- distilled water x μl (to 10 μl final volume)

incubate at 37°C for 1–2 h.

2 Heat inactivate the enzyme at 70°C for 15 min.

3 Add a threefold molar excess (60 pmol) of unlabeled complementary oligonucleotide, 1 μl 20× oligo annealing buffer and water to a combined volume of 20 μl.

4 Heat at 90°C for 5 min in a dri-block. Remove the block containing the reaction tube from the dri-block unit and allow it to cool slowly to room temperature to promote annealing.

5 Make the volume up to 100 μl with TE pH 7.6 and centrifuge through a Sephadex G-50 column to separate the radiolabeled oligonucleotide from unincorporated nucleotides. ④

6 Collect the eluate and determine the level of radioactivity in a scintillation counter. ⑤

Electrophoretic mobility shift assay

1 Assemble a 6% nondenaturing polyacrylamide gel in 0.5× TBE as follows. This is sufficient for a 16 cm × 16 cm × 1.5 mm gel.
- 40% acrylamide (29:1)   7.5 ml
- 10× TBE   2.5 ml

- distilled water x ml (to a final volume of 50 ml)
- 20% APS   150 µl
- TEMED   50 µl

overlayer the gel with distilled water and allow the gel to polymerize (15–20 min at room temperature).

2 Pre-run the gel for 1–2 h at 10–15 V/cm in $0.5\times$ TBE gel running buffer.

3 Set up the protein-DNA binding reaction as follows:
- nuclear extract ⑥   2–4 µg
- 5× binding buffer   4 µl
- poly (dI:dC)-poly(dI:dC) ⑦   0.5–2 µg
- ^{32}P-labeled DNA probe ⑧   1 µl
- distilled water x µl (to a final volume of 20 µl)

mix and incubate at room temperature for 15-30 min. ⑨⑩⑪⑫

4 Add 0.5 µl 10× EMSA loading buffer and apply samples to the polyacrylamide gel. Continue the electrophoresis until the free probe has traversed the length of the gel (for a 30-mer oligonucleotide and 6% acrylamide gel that is until the Bromophenol blue is 1–2 cm from the end of the gel).

5 Transfer the gel on to 3MM Whatman paper, cover gel with plastic film and dry under vacuum using a slab gel dryer.

6 Autoradiograph overnight at $-70°C$ with intensifying screens.

Protocol 1. Electrophoretic mobility shift assay

Protocol 2. Methylation interference analysis. K. Docherty

Reagents

40% (w/v) Acrylamide (see *Protocol 1*)

20% Ammonium persulfate (APS)

Chloroform/isoamyl alcohol (24:1)▽

Dimethyl sulfate (DMS)▽

Distilled water

DMS reaction buffer (0.1 mM EDTA, 10 mM $MgCl_2$, 50 mM sodium cacodylate pH 8.0)▽

DMS stop solution (2.5 M 2-mercaptoethanol, 3 M ammonium acetate pH 7.0)

5× EMSA binding buffer [125 mM Tris-HCl pH 7.5, 250 mM KCl, 5 mM DTT, 25% (v/v) glycerol] store at 4°C

10× EMSA gel loading buffer [50% (v/v) glycerol, 0.25% (w/v) xylene cyanol, 0.25% (w/v) Bromophenol blue]

70% Ethanol▽

Formamide dye mix (80% deionized formamide, 1× TBE, 1 mg/ml xylene cyanol FF, 1 mg/ml Bromophenol blue)

Gel elution buffer (10 mM Tris-HCl pH 7.5, 1 mM EDTA, 300 mM NaCl)

Nuclear extract (prepared as described in *Protocol 1*)

Phenol/chloroform/isoamyl alcohol (25:24:1) – saturated with 0.1 M Tris pH 8.0

Piperidine (supplied as a 10 M stock)▽

Poly (dI:dC)-poly(dI:dC) (4 mg/ml in 100 mM NaCl) store at −20°C.

3 M Sodium acetate pH 5.2

Radiolabeled oligonucleotide probe (prepared as described in *Protocol 1*)

10× TBE (108 g Tris-base, 55 g boric acid, 7.5 g EDTA) make up to 1 liter with deionized water, autoclave and store at room temperature

Equipment

Dri-block at 90°C

Fuji RX film

Ice bucket

Perspex radiation shield

Plastic film

Rotating wheel for microcentrifuge tubes

Speedvac concentrator

Vortex

Procedure

Methylation of the probe

1 Resuspend $1-10 \times 10^6$ d.p.m. of end-labeled, double-stranded oligonucleotide probe in 100 μl DMS reaction buffer in a 1.5 ml microcentrifuge tube. Add 1 μl of dimethyl sulfate (DMS) and incubate at room temperature for 5–10 min.

2 Stop the reaction by addition of 35 μl DMS stop solution.

3 Centrifuge through a Sephadex G-50 column and collect the eluate. ①

Electrophoretic mobility shift assay

1 Prepare an 8% nondenaturing polyacrylamide gel in 0.5× TBE as follows. This is sufficient for a 15 cm × 15 cm × 1.5 mm gel.
- 40% acrylamide   10 ml
- 10× TBE   2.5 ml
- distilled water   to 50 ml
- 20% APS   150 μl
- TEMED   75 μl

overlayer the gel with distilled water and allow the gel to polymerize (15–30 min).

2 Pre-run the gel for 30–60 min at 10–15 V/cm in 0.5× TBE gel running buffer.

3 To a 1.5 ml microcentrifuge tube add:
- nuclear protein extract x μl (10–100 μg)
- poly (dI.dC)   0.5–2 μg

13

Notes

Time required: 2 days

Day 1 – methylate probe, EMSA reactions, pour and run EMSA gel, identify free and protein bound bands, excise bands from gel, extract DNA from gel overnight.

Day 2 – cleave DNA with piperidine, wash DNA pellets to remove traces of piperidine, pour and run a denaturing polyacrylamide gel, carry out autoradiography.

① See *Protocol 1*.

② From Step 3.

③ For a 30 bp oligonucleotide and 8% acrylamide gel, i.e. until the Bromophenol blue is 1–2 cm from the end of the gel.

④ Freshly diluted from the 10 M stock.

Protocol 2. Methylation interference analysis

- 5× EMSA binding buffer   4 μl
- partially methylated DNA   2–10 ng (4–8 × 10^5 d.p.m.) ②
- distilled water   x μl (to final volume of 20 μl)

4 Incubate the reaction mixture at 20°C for 30 min.

5 Add 2 μl 10× EMSA gel loading buffer and apply samples to the polyacrylamide gel. Continue the electrophoresis until the free probe has traversed the length of the gel. ③

6 Wrap the gel in plastic film and expose to X-ray film for 1–30 min. Visualize DNA bands by autoradiography. Cut out gel slices containing retarded and free DNA.

7 Cut the gel slices into small pieces using a scalpel blade and place them in a 1.5 ml microcentrifuge tube. Add 300 μl gel elution buffer and place in a revolving wheel at room temperature for 3–4 h. Remove and save the buffer, add 300 μl fresh gel elution buffer and leave the tubes to rotate overnight at room temperature.

8 Pool the two samples of elution buffer in a 1.5 ml microcentrifuge tube and add 0.8 ml of phenol:chloroform:isoamyl alcohol. Vortex and centrifuge for 1 min in a microcentrifuge. Transfer the top layer to a fresh 1.5 ml microcentrifuge tube and add 1 ml of chloroform:isoamyl alcohol. Vortex and centrifuge for 1 min in a microcentrifuge. Transfer the top layer to a fresh 1.5 ml microcentrifuge tube and add 0.1 vol. of 3 M sodium acetate, pH 5.2 and 3 vols ethanol. Mix the solution and place in a dry ice methanol bath for 30 min. [1] Centrifuge for 15 min in a microcentrifuge and discard the supernatant. Add 1 ml cold 70% ethanol

to the pellet and leave the tubes in an ice bucket for 5 min. Centrifuge for 10 min in a microcentrifuge tube then carefully remove and discard the supernatant. Dry the pellet in a Speedvac™ concentrator or vacuum dessicator.

9 Resuspend the pellet in 100 μl 1 M piperidine. ④ Heat samples in closed tubes at 90°C for 30 min.

10 Lyophilize the samples in a vacuum dessicator or Speedvac™ concentrator. Add 100 μl distilled water and dissolve the contents of the tubes by vortexing for 30 sec. Lyophilize the samples as above, resuspend in 50 μl distilled water and lyophilize once more.

11 Resuspend dried samples in 2–4 μl of formamide dye mix. Heat to 90°C for 2 min in a dri-block, then transfer samples to ice.

12 Apply samples to a sequencing gel of appropriate composition to separate cleaved DNA fragments.

13 Dry gel and expose to X-ray film to give an autoradiograph.

Pause point

1 May be left at −20°C overnight.

Protocol 2. Methylation interference analysis

Protocol 3. **Southwestern analysis.** J. Philippe

Reagents

Acrylamide stock [33.5% (w/v) acrylamide, 0.3% (w/v) bisacrylamide]⚠

Ammonium persulfate 10% (w/v)

Binding buffer [250 mM Hepes pH 7.9, 30 mM $MgCl_2$, 500 mM KCl, 1 mM DTT poly (dI:dC)-poly (dI:dC)]

Blocking buffer (5% nonfat dry milk in 1× binding buffer)

Denaturation buffer (6 M guanidine hydrochloride in 1× binding buffer)

Denatured salmon sperm DNA①

Destaining solution [10% (v/v) methanol, 10% (v/v) acetic acid]

Distilled water

Gel fixer [50% (v/v) methanol, 10% (v/v) acetic acid]

Gel running buffer [25 mM Tris-base, 250 mM glycine, 0.1% (w/v) SDS]

Molecular weight protein markers

Oligonucleotides

$[\gamma^{32}P]$-ATP (3000 Ci/mmol, 111 TBq/mmol)⚠

Protein sample ②

Running gel buffer (1 M Tris-HCl pH 9.1)

2× Sample buffer [125 mM Tris-HCl pH 6.8, 4% (w/v) SDS, 20%

methanol, 10% (v/v) acetic acid]

Transfer tank buffer [25 mM Tris-base, 192 mM glycine, 20% (v/v) methanol]

T4 DNA ligase

T4 polynucleotide kinase

10× T4 polynucleotide kinase buffer – usually provided with the enzyme

10×T4 DNA ligase buffer – usually provided with the enzyme

TEMED

Equipment

Bench-top centrifuge

Cassette for autoradiograph

Heater

Ice bucket

Nitrocellulose membrane (Schleicher and Schuell)

Perspex radiation shield

Pipette

Power supply

Small glass receptacles

Shaker

Sponge pads

(v/v) glycerol, 1.44 M 2-mercaptoethanol, 0.1% Bromophenol blue]
Stacking gel [30% (w/v) acrylamide/0.44% (w/v) bisacrylamide]
Stacking gel buffer (0.5 M Tris-HCl pH 6.8)
Staining solution [0.2% (w/v) Coomassie blue, 50% (v/v)

Transfer apparatus
Radioisotope hood
Vertical slab gel electrophoresis unit
Water bath
3MM Whatman paper

Procedure

SDS polyacrylamide gel electrophoresis

1 Prepare a 10% polyacrylamide gel:acrylamide/bisacrylamide (33.5%/0.3%) with a 4% stacking gel. ③

2 Prepare samples for electrophoresis by mixing sample buffer with protein extracts (75–100 µg of nuclear extracts) in a 1:1 volume ratio. Include two additional samples containing molecular weight marker proteins in the same total volumes.

3 Heat samples at 100°C for 3 min in a dri-block. ④

4 Allow the samples to cool to room temperature and microcentrifuge at maximum speed for 5 min. Transfer the supernatant to a fresh tube. ⑤

5 Load the samples on to the polyacrylamide gel. ⑥

6 Apply a voltage of 8 V/cm while the sample migrates through the stacking gel and increase to 15 V/cm for the running gel. The run will take about 4 h and is completed when the marker dye reaches the bottom of the gel.

Notes

Time required: 1.5 days

Day 1 – (i) Prepare and run samples on the polyacrylamide SDS gel, set up transfer apparatus, transfer overnight; (ii) label the oligonucleotides, anneal them, set up ligation to run overnight.

Day 2 – Wash filter, subject filter-bound protein to denaturation/renaturation, wash filter, autoradiograph.

① Dissolve salmon sperm DNA in distilled water at 1 mg/ml, sonicate to reduce the size of the DNA to 200–300 bp in length, phenol extract and ethanol precipitate; dissolve in water, denature by boiling for 15 min, and rapidly chill on ice. Store DNA at 4°C.

② Whole cell, nuclear or cytoplasmic (S100) extract as described in *Protocol 1*.

③ A 10% gel with a ratio of acrylamide/bisacrylamide of 33/0.3 should resolve proteins of molecular weights 10–200 kDa.

④ During sample preparation, the 100°C step might cause

7 Remove the glass plates and mark the orientation of the gel by cutting its upper left corner.

Protein transfer

1 Place the gel in a plastic container, add approximately 30 ml transfer buffer and place the container on a rocking platform for 30 min. ⑦

2 Cut the nitrocellulose membrane to the dimensions of the gel and wet it by sliding it into the buffer. ⑧

3 Assemble the gel transfer sandwich with the cathode flat, the sponge pad, a sheet of 3MM Whatman paper (soaked in transfer buffer), the gel, the nitrocellulose membrane, a second sheet of Whatman paper, the second sponge pad and finally the anode.

4 Using a glass pipette as a roller, squeeze out any air bubbles between the nitrocellulose membrane and the gel and between the Whatman paper and nitrocellulose.

5 Place the transfer sandwich into the transfer chamber filled with transfer buffer.

6 Transfer at a constant voltage of 30 V (180 mA) overnight at 4°C with cooling and stirring. ⑨

7 After transfer, stain the gel and a nitrocellulose strip corresponding to one of the marker proteins lanes with Coomassie blue in order to check transfer efficiency. To do this fix the gel for 30 min in 100 ml of gel fixer,

⑤ some proteins to precipitate and be lost from the assay, in which case the sample can be incubated at 37°C for 10 min.

⑤ This step removes insoluble material and avoids protein 'streaking' during electrophoresis.

⑥ Samples should be about 20 μl. An equal volume of 1× sample buffer should be loaded in unused wells.

⑦ The gel size may change slightly due to the methanol.

⑧ Although nylon membranes, such as Zetaprobe™ have a higher capacity for protein binding than nitrocellulose, they give an unacceptably high background.

⑨ Efficient transfer of proteins is dependent on: (i) ratio of acrylamide to bisacrylamide; (ii) transfer time and voltage, – high voltages may cause overheating, therefore transfer at 4°C over relatively long periods; (iii) buffer strength – reduction in ionic strength may allow use of higher voltages; (iv) buffer type – elimination of methanol may increase transfer efficiency but may diminish binding of proteins to nitrocellulose.

⑩ Oligonucleotides representing each strand of a DNA control element with 5–10 surrounding nucleotides should be synthesized with cohesive ends.

⑪ See *Protocol 1*.

⑫ Analyze an aliquot on a nondenaturing polyacrylamide gel to check the efficiency of the ligation reaction. The DNA probe used to interact with proteins bound on the nitrocellulose membrane ideally should contain multiple copies (three to five) of the control element to enhance sensitivity.

then stain for 30 min in 100 ml of staining solution. Destain in 100 ml destaining solution (for 10 min) with two or three changes as the solution turns blue.

8 Rinse the filter in distilled water, dry at room temperature and save.

Radiolabeling the DNA

1 Set up the following reaction mix in a 1.5 ml microcentrifuge tube:
- 10× kinase buffer   2 µl
- sense oligonucleotide   10 ng⑩
- antisense oligonucleotide   10 ng⑩
- [γ-^{32}P]ATP (3000 Ci/mmol,   6 µl (60 µCi, 2.2 MBq)
  111 TBq/mmol)
- distilled water x µl (to 20 µl final volume)
- T4 polynucleotide kinase (10 U/µl)   1 µl.

2 Incubate at 37°C for 1 h.

3 Place reaction tube in a beaker containing hot water (90°C) and place the beaker in an ice bucket filled with ice. Leave for 2–3 hours until the water temperature reaches 20°C.

4 Add 2 µl 10× ligation buffer and 1 µl T4 DNA ligase (10 U/µl).

5 Incubate overnight at 15°C.

6 Purify catenated DNA from unincorporated radioactive ATP using a G50 Sephadex or spun column.⑪ Alternatively ethanol precipitation may be used.⑫

⑬ Renaturation of proteins is necessary for high affinity binding to sequence-specific DNA elements.

⑭ To assess specificity, membrane strips should be incubated with unlabeled specific and nonspecific DNA (catenated if catenated labeled probes are used). Therefore for each DNA probe, at least three lanes should be run. One nitrocellulose strip is incubated with the labeled DNA element and poly (dI:dC)-poly (dI:dC). Specific and nonspecific (or mutated) double-stranded oligonucleotides are added to the incubation mixtures of the second and third strips, respectively, in a 100-fold excess to the labeled probe.

Hybridization

1 Cut the membrane into strips corresponding to each lane of the gel and immerse each strip in 6 M guanidine HCl dissolved in $1\times$ binding buffer (20–40 ml) in a plastic container.

2 Place the plastic container on an orbital shaker and leave it to shake gently for 10 min.

3 Dilute with the same volume of $1\times$ binding buffer (this results in 3 M guanidine HCl) and continue shaking the container for 5 min.

4 Repeat the dilution step four additional times to give 1.5 M, 0.75 M, 0.38 M and 0.185 M guanidine HCl sequential dilutions, respectively. The container should be shaken for 5 min between each dilution.⑬

5 Wash the membranes in $1\times$ binding buffer for 10 min.

6 Place the membrane strips in blocking buffer (20–40 ml) and agitate them gently for 1 h.

7 Add the ^{32}P-labeled concatamerized oligonucleotide probe (5–10 $\times$ 10S c.p.m./ml) to $1\times$ binding buffer containing 0.25% nonfat dry milk and 10 μg/ml poly (dI:dC)-poly (dI:dC) (or 10 μg/ml salmon sperm DNA).⑭

8 Agitate the membrane strips gently in this solution at 4°C for 2 h.

9 Wash the membrane strips for 5–10 min in 100 ml $1\times$ binding buffer at 4°C. Repeat this washing step three times. Dry the membrane strips on Whatman paper for 1–2 min and autoradiograph for at least 12 h.

Protocol 4. **DNase I footprinting.** N. Savery, T. Belyaeva and S. Busby

Reagents

Absolute ethanol▽
10% Acetic acid/10% methanol▽
10% (w/v) Ammonium persulfate (freshly made)
10× Binding buffer (50 mM MgCl₂, 500 mM potassium glutamate, 10 mM DTT, 5 mg/ml bovine serum albumin, 200 mM Hepes pH 8.0)
Denaturing acrylamide mix▽ [6% acrylamide (acrylamide:bisacrylamide ratio 19:1), 7 M urea, 1× TBE buffer]
DNA fragment labeled specifically at one end with ^{32}P▽
DNase I (molecular biology quality)
DNase I dilution buffer (7.5 mM MgCl₂, 5 mM CaCl₂, 25 mM potassium glutamate, 62.5 mM KCl, 20 mM Hepes pH 8.0)
70% Ethanol▽
Gel loading buffer (40% deionized formamide,▽ ① 5 M urea, 5 mM NaOH, 1 mM EDTA, 0.025% Bromophenol blue, 0.025% xylene cyanol)
20 mg/ml Glycogen

Phenol/chloroform/isoamyl alcohol▽ (25:24:1 by vol, saturated with 0.1 M Tris-HCl pH 8.0)
Protein samples
Stop solution (0.3 M sodium acetate, 10 mM EDTA)
10× TBE (107.8 g Tris base, 55 g boric acid, 7.44 g EDTA) make up to 1 liter with deionized water, and store at room temperature
TEMED

Equipment

Autoradiography equipment (X-ray film, cassettes and developing facilities) or phosphorimager
Controlled temperature block
Gel dryer
Microcentrifuge and microcentrifuge tubes
Pipette plus tips
Sequencing gel apparatus (power supply, electrophoresis tank, glass plates, spacers and combs)
Vacuum desiccator
3MM Whatman paper

Procedure

1 Mix end-labeled DNA fragment (1–4 nM)② with the proteins to be footprinted③, in a total volume of 20 µl of 1× binding buffer. ④

2 Incubate the mixture at 37°C for 20–30 min.⑤

3 Add 3 µl of an appropriate dilution of DNase I directly into the sample. ⑥ Mix by stirring gently with the pipette tip for 15 sec. After an appropriate incubation time ⑦ stop the reaction by adding 200 µl of stop solution and mixing well. [1]

4 Extract with 200 µl of phenol/chloroform/isoamyl alcohol. Transfer 200 µl of the aqueous phase to a fresh microcentrifuge tube.

5 Extract again with 200 µl of phenol/chloroform/isoamyl alcohol. Transfer 190 µl of the aqueous phase to a fresh microcentrifuge tube.

6 Precipitate the DNA by adding 1 µl of 20 mg/ml glycogen and 400 µl of cold absolute ethanol. Mix, and incubate at −70°C for 15 min. [2]

7 Centrifuge for 15 min in a microcentrifuge at 4°C, and discard the supernatant.⑧

8 Add 600 µl of cold 70% ethanol. [2]

9 Centrifuge for 15 min in a microcentrifuge at 4°C, and discard the supernatant.

Notes

Time required: 3 days

Fragment preparation and labeling, preparation of sequencing ladder – approximately 2 working days. Footprinting reactions (12 samples) – approximately 3 h. Electrophoresis, fixing and drying gel – approximately 3 h. Autoradiography – overnight or longer.

① Stir 100 ml aliquots with 5 g of BioRad AG 501-XG8 resin, 20–50 mesh, for 1 h, then remove resin by filtration through a Whatman 3M filter; store tightly capped in 10 ml aliquots at −20°C.

② The DNA fragment should be purified from an acrylamide gel and labeled specifically at one end of one strand with ^{32}P (using T4 polynucleotide kinase or the Klenow fragment of DNA polymerase). The DNA should be of as high a quality as possible, and care should be taken not to introduce nicks into the DNA during the fragment preparation procedures.

③ A 'no protein' control reaction should always be included in each footprinting experiment, as only by comparison with such a control can the regions of DNA protected by a bound protein be detected.

④ Some experimentation may be necessary in order to find the optimal conditions. It is essential to ensure that any cofactor required for activity of the proteins under study is included in the binding reaction (for example the *E. coli* cyclic AMP receptor protein requires the presence of cAMP in order to bind its specific target on the DNA).

10 Dry the DNA pellet under vacuum. Resuspend the pellet in 8 μl of gel loading buffer. ⑨ ②

11 Incubate the samples at 90°C for 2 min, then load 4 μl of each sample on to a denaturing acrylamide gel. ⑩

12 When electrophoresis is complete fix the gel with 10% acetic acid/10% methanol and dry on to 3MM paper. The results can be visualized using autoradiography or a phosphorimager system.

⑤ The exact duration of this incubation is not critical, but should be sufficient for the protein–DNA complex to form. The temperature may also be altered.

⑥ A DNase I calibration should be performed whenever a new DNA fragment or different reaction conditions are used. Set up reactions which contain no added proteins and digest for 30 sec with 3 μl of a range of DNase I concentrations (5×10^{-4} to 5×10^{-3} Kunitz units per μl is a sensible starting range for most applications). The most appropriate concentration is that which leaves approximately 50% of the fragment undigested in this assay (this will ensure that most of the cut fragments will have been cut only once).

⑦ Incubation times for each experiment should be adjusted by trial and error to ensure that bands outside the footprint regions are equally intense in all lanes. Nonspecific competitor [e.g. 10 μg/ml poly d(AT) or sonicated salmon sperm DNA] may be included in the reaction mix to maintain a constant substrate concentration for the DNase I. Note that incubation times shorter than 30 sec are likely to be significantly affected by timing errors in the mixing process.

⑧ The presence of the radioactive DNA pellet can be checked by holding the bottom of the microcentrifuge tube to a Geiger counter. The absence of a radioactive pellet may be due to excessive DNase I digestion or contaminating nucleases.

⑨ It is important to ensure that the samples are fully

Protocol 4. DNase I footprinting

resuspended. This often requires that the samples are incubated in loading buffer for 10–15 min at 37°C before being vigorously vortexed.

(10) The samples are analyzed on a denaturing gel of the type typically used as sequencing gels. Mix 60 ml of denaturing acrylamide mix with 300 μl of 10% ammonium persulfate and 60 μl of TEMED. Pour between 35 cm × 42 cm glass plates held apart by 0.4 mm spacers, and insert a suitable comb. Allow the gel to polymerize for at least 1 h before use. The gels should be pre-run at 60 W constant power for at least 30 min before the samples are loaded (the electrophoresis buffer is 1× TBE). The exact percentage of the gel and the duration of electrophoresis varies according to the distance of the footprinted area from the labeled end of the DNA fragment; for a footprint beginning 40 bp away from the labeled end of the fragment a 6% acrylamide gel is suitable and electrophoresis should be stopped when the dark blue dye in the loading buffer is 3–5 cm from the bottom of the gel. A 'G + A' or 'G' ladder should be run in parallel with the footprinting samples to act as a marker. The ladders are prepared from the same labeled DNA fragment stock that is used for the footprinting experiments, using standard Maxam–Gilbert sequencing techniques (see *Protocol 7*).

Pause points

[1] May be left at room temperature until all samples are ready for Step 4.

[2] May be left at −20°C or −70°C.

Protocol 5. **Hydroxyl radical footprinting.** N. Savery, T. Belyaeva and S. Busby

Reagents

Absolute ethanol△

10% Acetic acid/10% methanol△

10% Ammonium persulfate (freshly made)

10× Binding buffer (50 mM $MgCl_2$, 500 mM potassium glutamate, 10 mM DTT, 5 mg/ml bovine serum albumin, 200 mM Hepes pH 8.0)

Denaturing acrylamide mix [6% acrylamide (acrylamide: bisacrylamide ratio 19:1), 7 M urea, 1× TBE buffer]△

DNA fragment labeled specifically at one end with ^{32}P (see *Protocol 4*)△

70% Ethanol△

Gel loading buffer [40% deionized formamide△ (see *Protocol 4*, Note 1), 5 M urea, 5 mM NaOH, 1 mM EDTA, 0.025% Bromophenol blue, 0.025% xylene cyanol]

20 mg/ml Glycogen

0.3% H_2O_2 (freshly diluted from 30% stock) ①

Iron/EDTA solution (freshly made – mix an equal volume of 0.2 mM ferrous ammonium sulfate and 0.4 mM EDTA)

Phenol/chloroform/isoamyl alcohol△ (25:24:1 by vol. saturated with 0.1 M Tris HCl pH 8.0)

Protein samples

Sodium ascorbate 10 mM

Stop solution (freshly made – mix 16 μl of 0.2 M EDTA pH 8.0 and 5 μl of 0.1 M thiourea per reaction)

10× TBE buffer (107.8 g Tris base, 55 g boric acid, 7.44 g EDTA; make up to 1 liter with deionized water, and store at room temperature

TEMED

Equipment

Autoradiography equipment (X-ray film, cassettes and developing facilities) or phosphorimager

Controlled temperature block

Gel dryer

Sequencing gel apparatus (power supply, electrophoresis tank, glass plates, spacers and combs)

Microcentrifuge and microcentrifuge tubes

Pipette plus tips

Vacuum desiccator

3MM Whatman paper

Procedure

1 Mix end-labeled DNA fragment (1–4 nM) ② with the proteins to be footprinted ③④, in a total volume of 35 µl of 1× binding buffer. ⑤

2 Incubate at 37°C for 20 min. ⑥

3 Place 5 µl of iron/EDTA solution, 5 µl of 0.3% H_2O_2 and 5 µl of sodium ascorbate solution on a clean, nonabsorbant surface (e.g. a petri dish). ⑦

4 Using a pipette set to 50–100 µl take up the iron/EDTA solution, the H_2O_2 and the sodium ascorbate solution in the same tip, each separated by a small air gap. Inject into the DNA protein mixture.

5 Incubate the reaction mixture at room temperature for 1 min.

6 Add 21 µl of stop solution and mix well. ▫1

7 Extract with 70 µl of phenol/chloroform/isoamyl alcohol. Transfer the aqueous phase to a fresh microcentrifuge tube.

8 Precipitate the DNA by adding 1 µl of 20 mg/ml glycogen and 150 µl of cold absolute ethanol. ▫2

9 Centrifuge for 15 min in a microcentrifuge at 4°C, and discard the supernatant. ⑧

10 Add 600 µl of cold 70% ethanol. ▫2

11 Centrifuge for 15 min in a microcentrifuge at 4°C, and discard the supernatant.

Notes

Time required: 3 days

Fragment preparation and labeling, preparation of sequencing ladder – approximately 2 working days. Footprinting reactions (12 samples) – approximately 4 h. Electrophoresis, fixing and drying gel – approximately 3 h.

① The water used in the preparation of all reagents should be of the highest available quality.

② The DNA should be of as high a quality as possible, and care should be taken not to introduce nicks into the DNA during the fragment preparation procedures. Labeled DNA fragments should be checked on a denaturing acrylamide gel before use. Nicks may also be introduced by nuclease contamination of the protein samples being analyzed.

③ A 'no protein' control reaction should always be included in each footprinting experiment.

④ Glycerol, a common component of protein storage buffers, inhibits the hydroxyl radical cleavage of DNA if present at a concentration greater than 0.5% (v/v).

⑤ It is essential to ensure that any cofactor required for activity of the proteins under study is included in the binding reaction (for example the *E. coli* cyclic AMP receptor protein requires the presence of cAMP in order to bind its specific target on the DNA).

⑥ The exact duration of this incubation is not critical, but should be sufficient for the protein–DNA complex to

12 Dry the DNA pellet under vacuum. [3]

13 Resuspend the pellet in 8 μl of gel loading buffer. ⑨ [2]

13 Incubate the samples at 90°C for 2 min, then load 3 μl of each sample on to a denaturing acrylamide gel. ⑩

14 When electrophoresis is complete the gel should be fixed in acetic acid/methanol mix and dried on to 3MM paper. The results can be visualized using autoradiography or a phosphorimager system.

form. The temperature may also be altered.

⑦ It is necessary to optimize the cleavage conditions to leave approximately 70% of the fragment uncut. Perform the cleavage reaction on naked DNA (i.e. no added proteins) and alter the concentration of the iron/EDTA and sodium ascorbate solutions until an appropriate level of DNA cleavage is observed (sodium ascorbate should always be present at several times the concentration of the iron/EDTA).

⑧ The presence of the radioactive DNA pellet can be checked by holding the bottom of the microcentrifuge tube to a Geiger counter. The absence of a radioactive pellet may be symptomatic of the action of contaminating nucleases.

⑨ It is important to ensure that the samples are fully resuspended. This often requires that the samples are incubated in loading buffer for 10–15 min at 37°C before being vortexed vigorously.

⑩ The samples are analyzed on a denaturing gel of the type typically used as sequencing gels (see *Protocol 4*).

Pause points

[1] May be left at room temperature until all samples are ready for Step 7.

[2] May be left at −20°C or −70°C.

[3] May be left at −20°C or −70°C (this is the best point at which to store samples).

Protocol 5. Hydroxyl radical footprinting

Reagents

Absolute ethanol△

10% Acetic acid/10% methanol△

10% Ammonium persulfate (freshly made)

5× Binding buffer (25% glycerol, 0.5 M NaCl, 25 mM MgCl$_2$, 0.5 mM EDTA, 5 mM DTT, 0.25 mg/ml bovine serum albumin, 100 mM Tris-HCl pH 8.0)

10× Buffer for *Taq* polymerase, usually supplied with enzyme (typical composition 500 mM KCl, 15 mM MgCl$_2$, 1 mg/ml gelatin, 0.5% Tween 20, 100 mM Tris-HCl pH 8.3)

Denaturing acrylamide mix△ [6% acrylamide (acrylamide: bisacrylamide ratio 19:1), 7 M urea, 1 × TBE buffer]

Deoxyribonucleotide stock containing each dNTP at a final concentration of 2.5 mM

DNA sample (fragment or plasmid)

70% Ethanol△

Gel loading buffer [40% deionized formamide△ (see *Protocol 4*, Note 1), 5 M urea, 5 mM NaOH, 1 mM EDTA, 0.025% Bromophenol blue, 0.025% xylene cyanol]

20 mg/ml Glycogen

Protein samples

3 M sodium acetate pH 7.0

Sterile distilled water

Stop solution (3 M ammonium acetate, 0.1 mM EDTA, 1.5 M 2-mercaptoethanol)

Taq DNA polymerase

TEMED

TBE buffer (89 mM Tris-borate, 89 mM boric acid, 2 mM EDTA)

10× TBE (107.8 g Tris base, 55 g boric acid, 7.44 g EDTA) make up to 1 liter with deionized water, and store at room temperature

Equipment

Autoradiography equipment (X-ray film, cassettes and developing facilities) or phosphorimager.

Controlled temperature block

Gel dryer

Microcentrifuge and microcentrifuge tubes

Perspex radiation shield

Pipette plus tips

Thermal cycler for PCR

200 mM $KMnO_4$ (freshly prepared)
Mineral oil
Oligonucleotide primer end labeled with ^{32}P ⚠
Phenol/chloroform/isoamyl alcohol (25:24:1 by vol. saturated with
0.1 M Tris-HCl pH 8.0)

Sequencing gel apparatus (power supply, electrophoresis
tank, glass plates, spacers and combs)
Vacuum desiccator
3MM Whatman paper

Procedure

1 Mix the DNA (4–10 nM) ① with the proteins to be footprinted, in a total
volume of 20 μl of 1× binding buffer. ②

2 Incubate the mixture at 37°C for 30 min. ③

3 Add 1 μl of 200 mM $KMnO_4$ solution to the sample. Mix by stirring gently
with the pipette tip for 15 sec. ④

4 Incubate the mixture at 37°C for 4 min.

5 Add 50 μl of stop solution and mix well. [1]

6 Add 130 μl of TE buffer and extract with 200 μl of
phenol/chloroform/isoamyl alcohol. Transfer the aqueous phase to a
fresh microcentrifuge tube.

7 Precipitate the DNA by adding 1 μl of 20 mg/ml glycogen and 400 μl of
cold absolute ethanol. Mix, and incubate at −70°C for 15 min. [2]

8 Centrifuge for 15 min in a microcentrifuge at 4°C, and discard the
supernatant.

Notes

Time required: 1 day
Footprinting reactions (12 samples) – approximately 8 h.
Electrophoresis, fixing and drying gel – approximately 3 h.
Autoradiography – overnight or longer.

① The DNA sample may be either a purified fragment or an
intact plasmid. Nicks in the DNA will cause a high
background and care should be taken to avoid their
introduction during sample preparation.

② Some experimentation may be necessary in order to find
the optimal conditions.

③ The exact duration of this incubation is not critical, but
should be sufficient for the protein–DNA complex to
form. The temperature may also be altered.

④ $KMnO_4$ is present in excess in this reaction.

⑤ Primers should be designed to anneal 60–70 bp from the
area of interest and should be 15–20 bases in length to
ensure specificity. End label 50 pmol of primer using T4
polynucleotide kinase and [γ-^{32}P ATP] then dilute to
1 μM.

Protocol 6. Potassium permanganate footprinting

9 Add 600 µl of cold 70% ethanol. 2

10 Centrifuge for 15 min in a microcentrifuge at 4°C, and discard the supernatant.

11 Dry the DNA pellet under vacuum.

12 Redissolve the DNA pellet in 15 µl of sterile distilled water.

13 Mix the following components in a microcentrifuge tube:
- permanganate-treated DNA from Step 12   7.0 µl
- end-labeled primer (1µM)⑤   1.0 µl
- 10× *Taq* DNA polymerase buffer   5.0 µl
- 2.5 mM dNTPs   2.0 µl
- sterile distilled water   34.5 µl
- *Taq* DNA polymerase (5 U/µl)   0.5 µl.

14 Overlay with 30 µl of sterile mineral oil.

15 Subject the sample to 15 or more cycles of denaturation, annealing and extension using a thermal cycler.⑥ For a primer with a T_m of 55°C a suitable cycle profile is, 1× (94°C 3 min, 50°C 2 min, 72°C 3 min), 15× (94°C 1 min, 50°C 2 min, 72°C 1.5 min), 1× (94°C 1 min, 50°C 2 min, 72°C 10 min).

16 Remove the mineral oil from the sample.

⑥ The annealing temperature should be approximately 5°C below the T_m of the primer.

⑦ The presence of the radioactive DNA pellet can be checked by holding the bottom of the microfuge tube to a Geiger counter.

⑧ It is important to ensure that the samples are fully resuspended. This often requires that the samples are incubated in loading buffer for 10–15 min at 37°C before being vigorously vortexed.

⑨ The samples are analyzed on a denaturing gel of the type typically used as sequencing gels.

17 Add 150 μl of TE buffer and extract with 200 μl of phenol/chloroform/isoamyl alcohol. Transfer the aqueous phase to a fresh microcentrifuge tube.

18 Precipitate the DNA by adding 20 μl of 3 M sodium acetate and 440 μl of cold absolute ethanol. Incubate at −70°C for 15 min. ☐2

19 Centrifuge for 15 min in a microcentrifuge at 4°C, and discard the supernatant. ⑦

20 Add 600 μl of cold 70% ethanol. ☐2

21 Centrifuge for 15 min in a microcentrifuge at 4°C, and discard the supernatant.

22 Dry the DNA pellet under vacuum. ☐2

23 Resuspend the pellet in 8 μl of gel loading buffer. ⑧ ☐2

24 Incubate the samples at 90°C for 2 min, then load 4 μl of each sample on to a denaturing acrylamide gel. ⑨

25 When electrophoresis is complete the gel should be fixed in acetic acid/methanol mix and dried on to 3MM paper. The results can be visualized using autoradiography or a phosphorimager system.

Pause points

☐1 May be left at room temperature until all samples are ready for Step 6.

☐2 May be left at −20°C or −70°C.

Protocol 6. Potassium permanganate footprinting

Protocol 7. Preparation of 'G' ladder by DMS treatment and piperidine cleavage. N. Savery, T. Belyaeva and S. Busby

Reagents
(additional to those used in DNase I footprinting, see *Protocol 4*)

Dimethyl sulfate (DMS)⚠

DMS buffer (0.1 mM EDTA, 10 mM MgCl$_2$, 50 mM sodium cacodylate pH 8.0)⚠

DMS stop solution (2.5 M 2-mercaptoethanol, 3 M ammonium acetate pH 7.0)

Piperidine (supplied as a 10 M stock)⚠

Sephadex G-50

3 M Sodium acetate pH 7.0

Equipment
(additional to that used in DNase I footprinting, see *Protocol 4*)

Fume cupboard

Bench-top centrifuge capable of spinning 50 ml centrifuge tubes at 700 g

50 ml Centrifuge tubes

Siliconized polymer or glass wool

Procedure

1 Mix 2–5 μl of ~400 nmol end-labeled DNA with 200 μl of DMS buffer in a microcentrifuge tube.

2 Add 1 μl of DMS. Mix, and incubate at room temperature for 90 sec.①

3 Add 50 μl DMS stop solution and mix well.

4 Precipitate the DNA by adding 1 μl 20 mg/ml glycogen and 800 μl of cold absolute ethanol. Mix, and incubate at −70°C for 15 min.

Notes

Time required: 2.5 hours

① All operations involving DMS or piperidine should be performed in a fume cupboard.

② 40 μl is required for each cleavage reaction, the 1 M stock should be prepared immediately prior to use.

③ See *Protocol 1.*

5 Centrifuge for 15 min in a microcentrifuge at 4°C, and discard the supernatant.

6 Add 600 μl of cold 70% ethanol.

7 Centrifuge for 15 min in a microcentrifuge at 4°C, and discard the supernatant. Dry the DNA pellet under vacuum.

8 Prepare a 1:10 dilution of the stock 10 M piperidine. ②

9 Resuspend the dried pellet in 40 μl of 1 M piperidine.

10 Incubate at 90°C for 30 min.

11 Pass the sample through a Sephadex G-50 spin column. ③

12 Add 4 μl 3 M sodium acetate pH 7.0, 88 μl cold absolute ethanol and 1 μl 20 mg/ml glycogen. Incubate at −70°C for 15 min.

13 Centrifuge for 15 min in a microcentrifuge at 4°C, and discard the supernatant.

14 Add 600 μl of cold 70% ethanol.

15 Centrifuge for 15 min in a microcentrifuge at 4°C, and discard the supernatant.

16 Dry the pellet under vacuum.

17 Resuspend pellet in 16 μl gel loading buffer. Load 0.5–2 μl as a calibration lane on a footprinting gel.

Protocol 7. Preparation of 'G' ladder

II *IN VIVO* FOOTPRINTING.   M.C.U. Hammond-Kosack

Introduction

Gene expression is regulated by the interaction of a wide variety of proteins with specific DNA sequences, and the analysis of these interactions is of critical importance. These interactions can be studied both *in vitro* and *in vivo*, but *in vivo* studies (*in vivo* DNA footprinting) offer the advantage of studying protein–DNA interactions in the context of intact, transcriptionally active chromatin. Also, it is not uncommon that binding sites detected *in vitro* are not occupied *in vivo*. For example, within the rat tyrosine aminotransferase (r TAT) gene a strong footprint was detected *in vitro* upon binding of the transcription factor CREBP, but this site was clearly unoccupied *in vivo* in a cell line expressing r TAT upon glucocorticoid stimulation [1]. This shows that *cis*-acting elements identified by *in vitro* assays may not interact with their cognate factors *in vivo*. Furthermore, using *in vivo* footprinting the dynamic interplay of transcription factors and promoter DNA can be studied by examining the occupancy of *cis* elements at different times after (hormone) stimulation or different stages of development.

Modification of DNA *in vivo*

In vivo DNA footprinting involves treating the cells or tissue with limiting amounts of reagents that can penetrate the nuclei and modify/cleave the chromosomal DNA. Nucleotides within the sequence of interest should react with the chemical as long as no proteins are bound, but should be refractory to chemical modification when a protein is bound tightly at a specific *cis*-acting element. The modified DNA is then extracted and cleaved specifically at modified bases and the reaction products are analyzed using purified DNA and DNA from *in vivo*-reacted samples in parallel. The absence of specific bands in the *in vivo* sample accordingly suggests that this nucleotide sequence was occupied by a protein at the time of reaction with the DNA-modifying chemical, thus rendering the nucleotides inaccessible for chemical modification. Using the ligation mediated PCR (LMPCR) methodology [2] under optimal conditions, less than $1\,\mu g$ chromosomal DNA, isolated from the reacted tissue, is sufficient for the analysis.

The procedure of LMPCR is outlined in *Figure 1* and involves five distinct steps, namely: (i) primer hybridization, (ii) extension, (iii) ligation, (iv) PCR amplification and (v) visualization of the PCR products by extension with an end-labeled primer. The specificity of the gene-specific primers is of critical importance because any cross-hybridization to similar sequences would obscure the banding patterns on the sequencing gel. After successful hybridization of the first gene-specific primer to the target sequence only, the conversion of the hybridized fragments into blunt ended ones determines the product yield, and thus the intensity of bands on the sequencing gel. The selection of primers deserves careful attention. As shown in *Figure 1b* there are principally two possible arrangements for the primers. The first gene-specific primer may or may not overlap the second, whereas the third primer must always overlap the position of the second one. Also the T_m ideally should increase, with the third primer having a higher T_m than the second. This is necessary to ensure preferential hybridization of the third, end-labeled primer after the PCR and thus efficient labeling of the PCR products. In order to test the chosen primers for their specificity prior to using LMPCR, PCR reactions should be performed using combinations of the gene-specific primers for both the top and bottom strands and ideally chromosomal DNA from the target tissue/cell line. It is not worth continuing with the chosen primers unless single bands of the expected sizes are obtained. Also, different PCR programmes should be performed with increasing annealing temperatures (up to 10°C above the calculated T_m) to determine the highest possible annealing temperatures for Steps 1, 4 and 5 (*Figure 1a*). The design of the double-stranded linker should include the following parameters: (i) the two primers should hybridize to form a unidirectional double-stranded linker with a 5′ overhang and a blunt 3′ end; (ii) the annealed linker should be stable at room temperature but denature easily at the temperatures used for PCR (Step 4); and (iii) the T_m of the longer linker primer should ideally be the same as that of the gene specific primers used during the PCR step.

Ligation mediated PCR

In vivo footprinting experiments should always include an *in vivo* control (DNA obtained from tissues or cells where the gene of interest is not expressed) as well as the *in vitro* reacted DNA, to ascertain that any observed footprints (i.e. missing bands) really indicate bound protein.

(a)

Base specific cleavage sites

1. Primer hybridization
2. Extension with sequenase

1st gene specific primer

Blunt ends generated at cleavage site

3. Ligation of universal linker with DNA ligase

Double-stranded linker

4. PCR cycling with *Taq* DNA polymerase, the linker primer and 2nd gene specific primer

2nd gene specific primer

Linker primer

25 ×

PCR products

5. Extension of gene-specific PCR products with end-labeled primer and *Taq* DNA polymerase

End-labeled 3rd gene specific primer

9 ×

Sequencing gel

a *in vitro* reacted protein-free DNA

b Chromosomal DNA from *in vivo* reacted tissue

a   b

(b)

$5'$ $3'$

$3'$ $5'$
$3'$ $5'$
$3'$ $5'$

$5'$ $3'$

$3'$ $5'$ $3'$ $5'$
$3'$ $5'$

$$T_m (\text{■■■}) < T_m (\text{▫▫▫}) < T_m (\text{▨▨▨})$$

$$T_m (\text{▭}) = T_m (\text{▫▫▫})$$

Figure 1. Schematic of the ligation mediated PCR (LMPCR) procedure (a) and possible arrangement of PCR primers (b). (a) Chemically modified and cleaved chromosomal DNA is hybridized with a first gene-specific primer and extended to yield a gene-specific blunt ended population of DNA representing all cleaved sites in the promoter (Steps 1 and 2). A unidirectional linker is ligated to the blunt ended fragments (Step 3), followed by PCR with a second gene-specific primer and a linker primer. This should amplify all cleaved DNA fragments within the promoter (Step 4). PCR products are visualized on a sequencing gel as a sequence ladder after extension with a third gene-specific, end-labeled primer (Step 5). Comparison between sequence ladders *in vitro* and *in vivo* modified DNA should identify foot-printed sequences (indicated here by the two boxes in lane b). (b) The T_m of the gene-specific primers should ideally increase from primer 1 to 3. Primer 2 and 3 should always overlap to allow for efficient labeling (Step 5) whereas primer 1 and 2 need not necessarily overlap.

Protocols provided

8. In vivo *footprinting*

References

1. Rigaud, G., Roux, J., Pictet, R. and Grange, T. (1991) *Cell*, **67**:977.
2. Mueller, P.R. and Wold, B. (1989) *Science*, **246**:780 .

Reagents

Annealed linker (Ib+II) – linker primer Ib (5'-GCAATCATTTGAGAGATCTGAATTC) and linker primer II (5'-GAATTCAGATC)

Bovine serum albumin (BSA)

$5\times$ Buffer 1 (200 mM Tris-HCl pH 7.5, 250 mM NaCl)

Buffer 2 (310 mM Tris-HCl pH 7.5)

n-Butanol⚠

Cells in culture

Chloroform

Culture medium

Dimethyl sulfate (DMS)⚠

Dissected tissue

DNA extraction buffer [10 mM Tris-HCl pH 9.5, 1% (w/v) SDS, 50 mM EGTA, 350 mM EDTA, 2 mg/ml proteinase K]①

dNTPs (2 mM)

70% Ethanol⚠

Oligonucleotide primers (1st or 2nd gene specific primers)

PCR buffer [100 mM Tris HCl pH 8.9, 500 mM KCl, 30 mM MgCl$_2$, 0.01% (w/v) gelatin]

Phenol

Phenol/chloroform 50–50% v/v

Phosphate buffered saline (PBS)

Piperidine (Kodak)

2-Propanol⚠

Sequencing loading buffer

3 M Sodium acetate (pH 5.2)

1% Sodium dodecyl sulfate (SDS)

Solution A (20 mM MgCl$_2$, 20 mM DTT, 200 μM dNTPs)

Solution B (17.5 mM MgCl$_2$, 42 mM DTT, 125 μg/ml BSA)

Solution C (14.7 mM MgCl$_2$, 29 mM DTT, 4.4 mM ATP, 74 μg/ml BSA)②

Sterile distilled water

T4 DNA ligase

T7 DNA polymerase

Taq DNA polymerase

TE buffer

Three gene specific primers for each DNA strand

Equipment

Agarose gel

Bench-top centrifuge

15 ml Centrifuge tubes

Microcentrifuge and microcentrifuge tubes

PCR thermal cycler plus tubes

Plastic spatula

6% Sequencing gel

Procedure

Treatment with dimethyl sulfate (DMS) ③

A. *For* in vivo *methylation*

1 Remove medium from the cell monolayer and replace it with fresh medium containing appropriate DMS concentrations.

2 Incubate for 5 min at room temperature. Remove the medium containing DMS and wash the cell monolayer several times with 5 ml fresh medium without DMS.

3 Add 2 ml PBS and, using a plastic spatula, scrape the cells off the dish. Transfer the cells to a 15 ml centrifuge tube and centrifuge on a bench-top centrifuge at 400 *g* for 5 min.

4 To prepare chromosomal DNA④ add 1× 5 ml extraction buffer, wetting the tissue/cells thoroughly and incubate at 55°C overnight. Centrifuge for 5 min in a microcentrifuge to pellet cell debris, and extract up to five times with phenol.⑤ Extract once with phenol-chloroform and chloroform. Take off the DNA phase,⑥ dilute threefold with sterile water and precipitate by adding 0.1 vol. 3 M sodium acetate pH 5.2, and 1 vol. 2-propanol. Mix well and centrifuge immediately for 5 min in a microcentrifuge. Resuspend the pellet in 0.5 ml TE and reprecipitate. Wash pellet with 70% ethanol, add 100 µl TE and allow it to resuspend overnight at 4°C. Assess DNA quality on an agarose gel.⑦

5 Dissected tissues are treated in the same manner as cultured cells, but in

Notes

Time required: 5 days

Day 1 – Treatment of DNA/cells/tissue with DMS <2 h; isolation of modified chromosomal DNA overnight.
Day 2 – Resuspension of chromosomal DNA >5 h; restriction enzyme digest; (to reduce viscosity, optional) overnight.
Day 3 – LMPCR (start mid-afternoon); Step 1, Primer hybridization, <45 min; Step 2, Strand extension, <30 min; Step 3, Ligation, overnight.
Day 4 – Step 4, Polymerase chain reaction, <4 h; Labeling of 3rd gene-specific primer <45 min; Step 5, Extension of PCR products, <2 h; Cast 6% sequencing gel, <30 min; Precipitation of labeled LMPCR products, overnight.
Day 5 – Resuspension of DNA in sequencing loading buffer, <30 min; Sequencing gel electrophoresis, 2–6 h; Autoradiography, 2 to several days.

① Make up fresh prior to use and use a freshly prepared proteinase K stock solution.

② Buffers 1 and 2 and solutions A–C can be stored in small aliquots at −20°C. Solutions A–C should not be thawed more than once.

③ The most commonly used chemical for *in vivo* footprinting is dimethyl sulfate (DMS). DMS modifies predominantly guanine residues, but adenine residues are also modified, although only to a minor extent.

④ Chromosomal DNA can be prepared by any method that

order to prepare the chromosomal DNA the tissue is frozen in liquid nitrogen and ground to a fine powder before adding the extraction buffer.

B. For in vitro *methylation*

1 Resuspend 10–20 µg chromosomal DNA⑧ in 100 µl distilled water in a microcentrifuge tube. Add 1 µl of DMS stock solution or appropriate dilutions thereof (e.g. 10%, 1%, etc.). Incubate the mixture at room temperature for 5 min.

2 Stop the reaction by adding 1 µl butanol (not water-saturated). Mix well by vortexing and centrifuge for 1 min in a microcentrifuge.⑨

3 Resuspend the pellet in 150 µl distilled water, add butanol to the top of the tube, mix well, centrifuge and dry the precipitated DNA *in vacuo* for at least 10 min. Add 150 µl 1 M piperidine⑩ and incubate the samples at 90°C for 30 min. Butanol-precipitate all samples, resuspend them in 150 µl 1% SDS, reprecipitate, resuspend them in 150 µl distilled water and precipitate with butanol for a third time. Dry the chemically modified DNA samples *in vacuo* for at least 30 min and resuspend at 0.2–0.5 µg/µl in distilled water.⑪ Subject samples to ligation mediated PCR in order to visualize the gene specific sequencing ladders.

Ligation mediated PCR (LMPCR)

All reactions can be performed in a PCR thermal cycler.⑫

⑤ As the extraction buffer is denser than phenol, the phenol phase will be the upper phase.

yields high molecular weight DNA without shearing.

⑥ The DNA now partitions in the upper phase.

⑦ Chromosomal DNA should migrate similarly to the λ DNA run as a marker.

⑧ Chromosomal DNA samples should be digested with a restriction enzyme (which however, should not cut within the region of interest) to reduce the viscosity prior to *in vitro* modification or piperidine cleavage.

⑨ A pellet should be visible at the bottom of the tube.

⑩ Dilute the piperidine in water from a 10–11 M stock solution.

⑪ Allow several hours, or overnight, for the cleaved DNA to dissolve fully.

⑫ When using a PCR thermal cycler with a heated lid, it is important to leave the lid off during the strand-extension and ligation steps (Steps 2 and 3) to avoid (partial) heat inactivation.

⑬ The annealing temperature should be close to the T_{m}.

⑭ The buffer/enzyme mix should be added immediately after the machine has cooled down to 20°C, by pipetting through the mineral oil and mixing by pipetting up and down three times for each sample. This seems to be one of the critical steps where speed is important.

⑮ The background can be considerably reduced by using any one of several 'hot-start' methods for PCR. In this case the primers are added after the initial denaturation.

Primer hybridization

1 In a PCR tube, add the following:
- chromosomal DNA x μl (0.5–5 μg)
- 1st gene specific primer (0.3 μM)   1 μl
- 5× buffer 1   3 μl
- distilled water x μl (to a final volume of 15 μl).

2 Place the tube in the PCR thermal cycler and set programe A: 95°C for 2 min, 50–60°C for 30 min, 20°C, hold. ⑬

Strand extension

1 Add 8.5 μl solution A, 0.5 μl T7 DNA polymerase (10 U/μl) and incubate the reaction mixture at 40°C for 10 min. ⑭

2 Stop the reaction by incubating at 65°C for 10 min and transfer to 15°C. This is achieved using programe B: 40°C for 10 min, 65°C 10 min, 15°C, hold.

Ligation

1 Prepare the ligation mixture by combining the following:
- solution C   17 μl
- annealed linker (Ib + II)   5 μl
- T4 DNA ligase (1 U/μl)   3 μl.

2 Combine the following:
- buffer 2   6 μl
- solution B   20 μl
- ligation mixture   25 μl.

⑯ The PCR products (5 μl aliquots) should be electrophoresed on 2% agarose gels. A smear over the range from ~30–500 bp should be observed. If the smear extends from the wells to the bottom, this is a sign of unspecific amplification, in which case the experiment should be aborted.

⑰ End-label 50 pmol of primer by T4 polynucleotide kinase using 10 μl γ-[^{32}P]-ATP in a volume of 50 μl. After Sephadex G-50 spin-column purification (see *Protocol 1*), 2.5 pmol of end-labeled primer (2.5 μl) are used for each extension.

⑱ The annealing temperature should be higher (~5–10°C) than during the PCR amplification, but not more than 10°C higher than the T_m of the third gene-specific primer to ensure efficient labeling of all PCR products.

⑲ In the initial tests it is advisable not to run the end-labeled primer off the gel. A clear gap of 25 nucleotides (corresponding to the length of the linker primer) should be present before the bands corresponding to G-residues within the promoter sequence start.

3 Add this combined 51 μl to the tube in the PCR thermal cycler and start
   programe C: 15°C for 12–18 h, 70°C for 10 min, −4°C hold.

4 Precipitate DNA with 10 μg glycogen as carrier by adding 0.1 vol. 3 M
   sodium acetate and 1 vol. 2-propanol. Mix well and centrifuge for 10 min
   in a microcentrifuge. Rinse the pellet in 70% ethanol and redissolve DNA
   pellet in 20 μl distilled water.

Polymerase chain reaction

1 Add the following to a microcentrifuge tube:
   - DNA (from Step 4, Ligation)   20.0 μl
   - 10× PCR buffer   5.0 μl
   - dNTPs (2 mM)   5.0 μl
   - *Taq* DNA polymerase (5 U/μl)   0.5 μl
   - distilled water   16.5 μl.

2 Heat at 94°C for 2.5 min.

3 Initiate the PCR cycling programe,   pausing after the denaturation step
   (40 sec at 94°C) during the first cycle to add 3 μl of primer mix (first or
   second gene-specific primer and linker Ib). Continue cycling using
   programe D: (94°C for 40 sec, 58–72°C for 1 min, 76°C for 1 min) ×25,
   −4°C hold.⑮

4 Use 25 μl of the PCR products for the final extension, store the remainder
   at −20°C.⑯

Labeling

1 Combine the following in a microcentrifuge tube:
   - PCR products (Step 4, PCR)   25.0 μl
   - dNTPs (2 mM)   5.0 μl
   - 10× PCR buffer   2.5 μl
   - *Taq* DNA polymerase (5 U/μl)   0.5 μl
   - end labeled second or third gene-specific primer ⑰   2.5 μl
   - water   14.5 μl.

2 Extend the gene-specific PCR sequence ladder using programe E:
   94°C for 2.5 min (94°C for 40 sec , 59–76°C for 3 min, 76°C for 5 min)
   ×9, −4°C hold. ⑱

3 Phenol-chloroform extract samples and precipitate with 0.1 vol. 3 M
   sodium acetate and 1 vol. 2-propanol. Centrifuge for 10 min in a
   microcentrifuge.

4 Wash the pellets thoroughly in 70% ethanol, briefly air dry samples and
   redissolve in 10 μl sequencing loading buffer.

5 Load 1–2 μl on a 6% sequencing gel and electrophorese at 35 W until the
   free end-labeled primer runs off the bottom (5–10 min after the
   Bromophenol blue has run off). ⑲

6 Expose the fixed and dried gel to autoradiographic film for between 2 h
   and 5 days depending on signal strength.

III CLONING OF TRANSCRIPTION FACTORS. M.D. Walker

Introduction

Many of the techniques developed for analyzing the properties of transcription factors can be applied to partially purified or even crude nuclear extracts. However, in order to obtain detailed knowledge of the structure and function of a transcription factor, it is essential to isolate cDNA clones encoding the protein. With such clones in hand, one can determine the amino acid sequence of the protein (inferred from the cDNA sequence), study the expression of the gene at the RNA and protein level, perform functional analysis of the protein by site-directed mutagenesis, and ultimately define structure and function through X-ray crystallography and targeted mutagenesis of the gene *in vivo* by generating mutant 'knockout' mice.

Cloning of mammalian transcription factors is typically accomplished through one of two major routes; the biochemical approach, and the direct cloning approach. In the first, protein is purified from an appropriate source (cultured cells or tissue) using conventional biochemical procedures; this typically involves a step of affinity chromatography [1] using a matrix containing immobilized DNA of appropriate sequence. More recently, epitope-tagged proteins have been used to facilitate purification of subunits of multimeric transcription factors [2,3]. Amino acid sequence information is obtained from the purified material and this is used to design oligonucleotides. These, in turn, are employed to isolate cDNA clones either by hybridization with a library or via PCR amplification of cDNA. In the direct cloning approach, expression cDNA libraries are screened using an appropriate probe, either recognition site DNA sequence or an interacting protein.

Purification of transcription factors by affinity chromatography

This approach relies on the ability of sequence-specific DNA-binding proteins to recognize their target sequence, and bind to it with relatively high affinity, in the presence of excess nonspecific DNA. The method involves the incubation of a partially purified protein extract with a large excess of soluble, competitor DNA, followed by passing it over a DNA-affinity column constructed with the target DNA. The proteins of interest partition to the DNA bound to the column, while the other DNA-binding proteins flow through, bound to the soluble nonspecific DNA. A key issue when adopting this approach is the relative abundance of the specific protein. Transcription factors are typically low abundance proteins, in the range of 10^3–10^5 molecules per cell [4]. Accumulation of sufficient purified protein to obtain sequence information (10–50 pmol of purified protein) usually demands large amounts of starting material (>10^{10} cells). For large organs or cultured cells adapted to grow in spinner culture such amounts may be easily obtainable. The amount of material needed will also depend on the quality of the sequencing facilities available to the researcher. As techniques improve, less protein will become necessary for obtaining sequence information.

Plaque and colony hybridization using degenerate oligonucleotide probes

Because of the redundancy of the genetic code, it is impossible to predict, unambiguously, the nucleotide sequence corresponding to a particular peptide. In fact only methionine and tryptophan, two of the rarest amino acids, are specified by a single codon, whereas the common amino acids leucine, serine and arginine are each specified by six codons. Two types of strategy have been developed to deal with this problem; use of short, fully degenerate oligonucleotides, and use of longer, minimally degenerate oligonucleotides. In the first, an appropriate peptide (containing a minimal number of highly degenerate amino acids) is used to design a fully degenerate mixed probe containing all possible combinations of oligonucleotides. The advantage is that included in this collection will be the 'correct' sequence; the disadvantage is that this 'correct' sequence will constitute a small percentage of the total probe, tending to compromise the resulting signal:noise ratio. In the second approach, a longer oligonucleotide 'guessmer' is synthesized with a minimally redundant or unique sequence. The sequence design is based on a variety of information including codon usage tables and the relative under-representation of CpG in mammalian genomes. When these factors are taken into consideration it can be estimated that the average homology of a guessmer to its correct sequence is ~85% [5]. Hybridization conditions are chosen such that imperfect hybrids can form without unacceptably high background signal from irrelevant sequences. An additional method for reducing the degeneracy of oligonucleotide probes is the use of a neutral base at positions of doubt, for example inosine can base pair with A and C [6].

Cloning of transcription factors

The decision whether to prepare short, fully degenerate oligonucleotides or longer guessmers will depend, in part, on the length of peptide sequence available and the degeneracy of the sequence. In the event of short oligonucleotides being chosen, they may be used as primers for PCR reactions (using cDNA or phage/plasmid DNA from a library) or as hybridization probes to screen plasmid or phage libraries. Bacteriophage libraries are more commonly used because of their greater efficiency in construction and screening. A difficulty in using highly redundant oligonucleotide mixtures as hybridization probes stems from the fact that under standard conditions, G–C base pairs are more stable than A–T. Thus short stretches of G–C in some oligonucleotides in the pool may form hybrids with irrelevant sequences of similar stability as that formed between the correct oligonucleotide and the desired sequence. This can make isolation of the correct clone difficult or even impossible. The problem can be circumvented by the use of tetramethylammonium chloride (TMAC), in the presence of 3 M TMAC, G–C and A–T base pairs have equivalent stability. Thus in a mixture of oligonucleotides of the same length but different sequence, the T_m of all oligonucleotides falls in a narrow range, simplifying selection of a suitable hybridization temperature.

Cloning by PCR

In recent years cloning by PCR has become increasingly popular. The availability of oligonucleotide sequence to two regions of a protein permits design of primers for PCR amplification of cDNA or DNA prepared from a library. If sequence is available for only one region, one sided 'anchored' PCR can be employed using, as second primer, a sequence corresponding to flanking vector sequences [7] or a sequence ligated to the end of the cDNA [8]. The exceptional sensitivity of PCR permits the use of highly degenerate oligonucleotides. However, mispriming events can lead to problems of false positives, and experiments should be designed carefully to try to minimize this problem [9].

Screening expression libraries (Southwestern)

In many cases, biochemical purification is not practical and direct cloning will be the method of choice. If the desired gene belongs to a known class of transcription factor (e.g. bZIP or bHLH) it may be possible to screen libraries using a protein probe or a redundant oligonucleotide designed according to a consensus sequence [10]. Often however, the only information available, through mobility shift analysis or footprinting, is the DNA-binding specificity of the protein, hence the popularity of the Southwestern screening procedure. In the original description of this procedure [11], the radioactive probe contained a single copy of the binding site. Subsequently, procedures have emphasized the improved signals produced by concatenated binding sites [12,13]. For some DNA-binding proteins, a cycle of denaturation–renaturation is advantageous for producing efficient binding [12]. Presumably the treatment can renature misfolded proteins. Although success with this procedure does not require that cDNA clones contain the entire protein-coding sequence, obviously an intact DNA-binding domain must be present. In cases where this domain is encoded by sequences located towards the 5′ end of a long RNA, it may be extremely difficult to clone by expression screening. Clearly, the quality of the library can be a decisive factor. To increase chances of success, the library should be derived from a cell type expressing highest levels of the transcript of the desired gene. Priming of cDNA using a combination of oligo-dT and random primers should be considered in order to improve representation of sequences residing at the 5′ end of long transcripts. If a commercial library is used, one should be aware of recent reports describing heavy contamination of certain mammalian libraries with yeast [14] and bacterial [15] sequences. Cloned sequences should be tested to verify their presence in mRNA or genomic DNA of the appropriate species.

Obtaining candidate clones using the Southwestern procedure is usually much faster and cheaper than the biochemical approach. However, it should be borne in mind that separating real positives from false positives can be quite time consuming. Following plaque purification, candidate clones can be tested with mutant binding site probes to evaluate their significance. Antibodies directed against cloned proteins can be generated and used in mobility shift experiments to test the relationship with a particular DNA binding activity. Finally, it must be remembered that many transcription factors cannot be cloned through the Southwestern procedure. Notable examples include transcription factors which bind DNA only as heteromeric complexes, or following a specific post-translational modification.

Protocols provided

9. *Purifying transcription factors by DNA affinity chromatography*
10. *Plaque and colony hybridization using degenerate oligonucleotide probes*
11. *PCR cloning of transcription factors using degenerate primers*
12. *Screening expression cDNA libraries*

References

1. Kadonaga, J.T. and Tjian, R. (1986) *Proc. Natl Acad. Sci. USA*, **83**:5889.
2. Field, J., Nikawa, J., Broek, D., MacDonald, B., Rodgers, L., Wilson, I.A., Lerner, R.A. and Wigler, M. (1988) *Mol. Cell. Biol.* **8**:2159.
3. Zhou, Q., Lieberman, P.M., Boyer, T.G. and Berk, A.J. (1992) *Genes Dev.* **6**:1964.
4. Nicolas, R. H. and Goodwin, G. H. (1993) in *Transcription Factors: a Practical Approach* (D.S.Latchman, ed.). IRL Press, Oxford.
5. Lathe, R. (1985) *J. Mol. Biol.* **183**:1.
6. Sambrook, J., Fritsch, E.F. and Maniatis, T. (1989) *Molecular Cloning: a Laboratory Manual.* Cold Spring Harbor Laboratory Press, Cold Spring Harbor, NY.
7. Ohlsson, H., Karlsson, K. and Edlund, T. (1993) *EMBO J.* **12**:4251.
8. Frohman, M.A., Dush, M.K. and Martin, G.R. (1988) *Proc. Natl Acad. Sci. USA,* **85**:8998.
9. Ashworth, A. (1993) in *Transcription Factors: a Practical Approach* (D.S.Latchman, ed.), pp. 125–142. IRL Press, Oxford.
10. Benezra, R., Davis, R.L., Lockshon, D., Turner, D.L. and Weintraub, H. (1990) *Cell,* **61**:49.
11. Singh, H., LeBowitz, J.H., Baldwin, A.S., Jr and Sharp, P.A. (1988) *Cell,* **52**:415.
12. Vinson, C.R., LaMarco, K.L., Johnson, P.F., Landschulz, W.H. and McKnight, S.L. (1988) *Genes Dev.* **2**:801.

13. Singh, H. (1993) *Methods Enzymol.* **218**:551.
14. Lovett, M., Kere, J. and Hinton, L.M. (1991) *Proc. Natl Acad. Sci. USA*, **88**:9628.
15. Dean, M. and Allikments, R. (1995) *Am. J. Hum. Genet.* **57**:1254.
16. Kerrigan L.A. and Kadonaga J.T. (1993) in *Current Protocols in Molecular Biology* (F.M. Ausubel *et al.*, eds), pp. 12.10.1–12.10.18. Greene/Wiley-Interscience, New York.
17. Knopf, J.L., Lee, M.H., Sultzman, L.A., Kriz, R.W., Loomis, C.R., Hewick, R.M. and Bell, R.M. (1986) *Cell*, **46**:491–502.
18. Wood, W.I., Gitschier, J., Lasky, L.A. and Lawn, R.M. (1985) *Proc. Natl Acad. Sci. USA*, **82**:1585–1588.
19. Jacobs, K.A., Rudersdorf, R., Neill, S.D., Dougherty, J.P., Brown, E.L. and Fritsch, E.F. (1988) *Nucleic Acids Res.* **16**:4637–4650.
20. Duby, A., Jacobs, K.A. and Celeste, A. (1994) in *Current Protocols in Molecular Biology* (F.M. Ausubel *et al.*, eds), pp. 6.4.3–6.4.10.
21. Marchuk, D., Drumm, M., Saulino, A., Collins, F.S. (1991) *Nucleic Acids Res.* **19**:1154.

Protocol 9. **Purifying transcription factors by DNA affinity chromatography.**①
A. Admon

Reagents

10 M Ammonium acetate

Annealing buffer (TE with 50 mM Tris, 10 mM $MgCl_2$ pH 7.6)

2-Butanol△

Cell washing buffer (PBS containing 3 mM KCl and 1 mM $MgCl_2$)

Cell swelling buffer (10 mM Tris-HCl pH 7.9, 10 mM KCl, 1.5 mM $MgCl_2$ with freshly added 1 mM DTT, 1 mM PMSF and 1 mM sodium metabisulfite)

Chloroform

Column storage buffer [0.3 M NaCl, 10 mM Tris-HCl pH 7.5, 1 mM EDTA, 0.02% (w/v) sodium azide]

Column regeneration buffer (2.5 M NaCl, 10 mM Tris-HCl pH 7.5 and 1 mM EDTA)

Competitor DNA②

CNBr-activated agarose

Distilled water

70% Ethanol△

1 M Ethanolamine-HCl pH 8.0

Formamide gel loading buffer (98% deionized formamide, 10 mM EDTA pH 8.0, 0.025% xylene cyanol FF, 0.0025% Bromophenol blue)

Gel filtration buffer [50 mM Tris-HCl or Hepes-KOH pH 7.9,

(fresh), 5 μCi [γ^{32}P] ATP△]

4 M LiCl

Ligase buffer (70 mM Tris-HCl pH 7.5, 2 mM ATP, 2 mM DTT)

1 M $MgCl_2$

Nonionic detergent such as NP-40 or LDAO

Nuclei extraction buffer [50 mM Tris-HCl pH 7.5, 20% (v/v) glycerol, 10% (w/v) sucrose, 0.42 M KCl, 5 mM $MgCl_2$ 0.1 mM EDTA, with freshly added 2 mM DTT, 1 mM PMSF, 1 mM sodium metabisulfite]

Phenol/chloroform 1:1 (v/v)

1 M Potassium phosphate pH 8.0.

2-Propanol

Synthetic target oligonucleotides

T4 DNA ligase

T4 Polynucleotide kinase

TEN buffer (10 mM Tris-Cl pH 7.5, 1 mM EDTA and 100 mM NaCl)

Equipment

Bench-top centrifuge

Dounce homogenizer with loose B pestle

EconoColumn (BioRad)

Equipment and reagents for DNA footprinting or gel shift assays

12.5 mM MgCl$_2$, 1 mM EDTA, 100 mM KCl, 20% (v/v) glycerol, 1 mM DTT (freshly added)].
Kinase buffer [50 mM Tris-HCl pH 7.5, 10 mM MgCl$_2$, 1 mM spermine, 1 mM EDTA, 3 mM ATP pH 7.0, 3 mM DTT

Equipment and reagents for SDS–PAGE
50 ml Sinter-glass funnel
TLC plates with fluorescent indicator or an intensifying screen
1.5 mm, 16% Polyacrylamide gel with 8.0 M urea and 1× TBE

Procedure

Purification of oligonucleotides

1 Design and synthesize oligonucleotides of 14–40 bases to form one or more protein-binding sites per double-stranded oligonucleotide, leaving at least two nucleotides between each *cis* element and the ends of the oligonucleotide. The oligonucleotides should also have four overhanging bases at the 5′ end to allow efficient annealing and polymerization to form long pieces of DNA of a few hundred nucleotides with multiple binding sites on each.

2 Dilute 1 μmol of the synthesized oligonucleotide (usually in ammonium hydroxide) fourfold in water and lyophilize to dryness.

3 Dissolve the dry oligonucleotides in formamide gel loading buffer, heat to 95°C for 3 min and load on a 1.5 mm, 16% polyacrylamide gel containing 8.0 M urea and 1× TBE (see *Protocol 4*).

4 Following electrophoresis, visualize the oligonucleotides by UV shadowing over an intensifying screen or TLC plates containing a fluorescent indicator. Excise the appropriate bands and extract the oligonucleotides by crushing the gel piece and shaking it overnight in 1 ml TE in 15 ml tubes.

Notes

Time required: approximately 5.5 days

Variable depending on purification steps used in the preparation of the nuclear extract.

① The different methods of DNA-affinity purification have been described in detail recently [16]. This protocol is compiled from laboratory manuals written by Jim Kadonaga, Dirk Bomann and Karen Perkins at the laboratory of Robert Tjian.

② Use double-stranded synthetic poly d(I–C), poly d(A–T) or natural DNAs from calf thymus, salmon sperm or *E. coli*. Anneal the synthetic DNAs by heating to 90°C for 10 min at a concentration of 10 mg/ml in TEN buffer and allow it to cool down slowly. Fragment the competitor DNAs by sonication to <1 kb and check fragmentation efficiency by agarose gel electrophoresis.

③ Check the efficiency of ligation by running a small fraction of the sample on a 2% agarose gel, the ligated DNA should be 500–1000 bp long. If the ligation has not

Protocol 9. Purifying transcription factors

5 Remove the acrylamide pieces by filtration, concentrate the oligonucleotides and remove the urea by sequential extraction with 2-butanol, each time add an equal volume of 2-butanol to the aqueous phase. When the aqueous phase shrinks to about 0.2–0.4 ml, transfer it to a microcentrifuge tube and precipitate the oligonucleotide by adding 0.4 M LiCl, 10 mM $MgCl_2$ and 2.5 vol. ethanol. Reprecipitate again, wash once with 70% ethanol and resuspend in 50 µl of TE. Quantify the DNA by UV absorbance at 260 nm assuming 1 OD_{260} = 40 µg/ml.

Preparation of the DNA affinity column

1 Mix equal molar amounts of each purified oligo (about 500 µg in TE) in annealing buffer and heat for 5 min at 90°C. Allow to cool slowly to room temperature.

2 Phosphorylate the 5′ end of the oligonucleotides in kinase buffer with T4 polynucleotide kinase for 2 h at 37°C. Stop the reaction by adding 1/10 vol. 4 M LiCl or 10 M ammonium acetate and heat to 65°C for 15 min.

3 Precipitate the oligonucleotides with 2.5 vols ethanol and wash once with 70% ethanol.

4 Redissolve the pellet in ligase buffer and add T4 DNA ligase. Incubate overnight at 15°C. ③

5 Extract once with phenol/chloroform and once with chloroform and reprecipitate by adding 1/3 vol. 10 M ammonium acetate and 1/3 vol. 2-propanol. Incubate for 20 min at −20°C and microcentrifuge for 15 min at 4°C. Wash the pellet with 70% ethanol and vacuum dry. Redissolve the DNA in distilled water (not TE). ④

succeeded, try different ligation temperatures, such as 4°C, room temperature or even 37°C.

④ The 2-propanol precipitation should remove unincorporated ATP which would otherwise compete with the oligonucleotides for the binding sites on the activated agarose.

⑤ For satisfactory coupling at least 90% of the radioactivity should remain on the column relative to the filtrate.

⑥ Before loading the extract on the column, the DNA binding activity is partially purified, usually by ammonium sulfate precipitation, followed by gel filtration on a Sephacryl-300 column in gel filtration buffer or by PVA precipitation followed by heparin chromatography. Purification of the DNA binding activity is followed by footprinting or gel shift assays (see *Protocol 1*).

⑦ For example 20 OD units (260 nm) of d(I–C) and 100 mg of calf thymus DNAs were successfully used for the first affinity pass for 120 ml of S-300 pooled fractions containing ~200 mg of nuclear extract protein.

6 Wash the CNBr-activated agarose three times with ice-cold distilled water, and once with ice-cold 10 mM potassium phosphate, pH 8.0, in a sinter-glass funnel. Do not allow the agarose to dry out at any stage. Gently transfer the resin to a 15 ml tube and add the ligated oligonucleotides resuspended in 4 ml of 10 mM potassium phosphate buffer, pH 8.0.

7 Allow the coupling reaction to proceed by rotating the tube overnight at room temperature. Transfer the agarose slurry to a sinter-glass funnel and wash it twice with 100 ml distilled water, collecting the first few milliliters of the filtrate. ⑤

8 Block the remaining active groups by treating the slurry with 100 ml of 1 M ethanolamine HCl, pH 8.0, and wash sequentially with 100 ml of 10 mM potassium phosphate, pH 8.0, 100 ml of distilled water and 100 ml column storage buffer. The agarose can be stored for at least a year in a capped tube in the storage buffer at 4°C.

Preparation of the nuclear extract

1 Harvest the cells in washing buffer and centrifuge at 1000 g for 4 min. Resuspend the cell pellet in cell-swelling buffer, 10 ml per 10^8 cells. Allow the cells to swell for 20 min on ice, then homogenize in a Dounce homogenizer using 20 strokes of a B-pestle. Centrifuge immediately at 2000 g at 4°C. Remove the cytoplasmic supernatant and store the nuclear pellet at $-20°C$.

Protocol 9. Purifying transcription factors

2 Extract the DNA-binding proteins from fresh or frozen nuclei by stirring them on ice for 30 min in nuclei extraction buffer. Clear the solution of debris by centrifugation at 50 000 g for 1 h at 4°C.

3 Separate nucleases and other interfering contaminants from the DNA-binding protein by using one or more purification steps.⑥

DNA affinity chromatography

1 Incubate the extract for 10 min on ice with the competitor DNA. ⑦

2 Centrifuge the extract at 10 000 g for 10 min at 4°C to remove newly formed aggregates that can clog the DNA-affinity column. Load the proteins with the competitor DNA on similar DNA-affinity columns in parallel, and connect any number of different DNA columns in series (to retain different DNA-binding proteins). This step should be performed at 4°C at a flow rate of not more than 15 ml/h, using 1 ml of resin in an EconoColumn.

3 Separate the series of columns and wash them with 10 vols of column buffer containing 0.1% nonionic detergent such as NP-40 or LDAO, omitting the PMSF and the metabisulfite. Elute the DNA-binding proteins with column buffer containing 0.6–1 M KCl and collect small fractions into microcentrifuge tubes.

4 Check for the presence of the desired proteins by footprinting or gel shift assays and check their purity by SDS-PAGE and silver staining.

5 In the event that another round of DNA-affinity column needs to be employed (with a slightly different DNA sequence), pool the appropriate fractions from the parallel columns, dilute them at least sixfold with the column running buffer without KCl to bring the salt concentration to 0.1 M, mix them with about 2–5% the amount of the competitor DNA and load on a single DNA-affinity column. Wash, elute and test for activity and for purity as before. Freeze small aliquots in liquid nitrogen and store the DNA binding proteins at $-70°C$ avoiding multiple thaws.

6 Regenerate the DNA affinity columns for reuse (many times) by washing with column regeneration buffer and store them in column storage buffer at $4°C$.

Reagents

- $[\gamma^{32}P]$ATP 3000 Ci/mol, 111 TBq/mmol ▽
- Bacteriophage or plasmid library
- 5× Denhardt's solution (0.1% Ficoll 400, 0.1% polyvinylpyrrolidone, 0.1% bovine serum albumin)
- Distilled water
- 10× Kinase buffer (0.5 M Tris-HCl pH 7.6, 0.1 M $MgCl_2$, 50 mM DTT, 1 mM spermidine HCl, 1 mM EDTA)
- Oligonucleotide ①
- Sephadex G50
- Sodium dodecyl sulfate (SDS)
- Sonicated, denatured salmon sperm DNA (SS DNA)
- SSC hybridization buffer (SHB) (6× SSC, 0.1% SDS, 100 μg/ml SS DNA, 5× Denhardt's solution)
- 6× Standard saline citrate (SSC) (0.9 M NaCl, 0.09 M sodium citrate pH 7.0)
- Tetramethylammonium chloride (TMAC)

- T4 polynucleotide kinase
- TMAC hybridization buffer (THB) (3 M TMAC, 0.05 M sodium phosphate pH 6.8, 1 mM EDTA, 5× Denhardt's, 0.6% SDS, 100 μg/ml SS DNA) ②
- TMAC wash buffer (TWB) (3 M TMAC, 0.05 M Tris-HCl pH 8.0, 0.2% SDS)

Equipment

- Bench-top centrifuge
- Centrifuge tubes, 15 ml polypropylene (for spun G50 column)
- Glass crystallization dishes
- Nitrocellulose filters 0.45 μm, 132 mm diameter
- Plastic film
- Sealable bags
- Shaking incubator
- Water baths

Procedure

A. SSC procedure [17]

1 Prepare duplicate nitrocellulose filters bearing bacteriophage plaques or bacterial colonies (up to 50 000/15 cm plate). ③

2 Wash filters three to five times for 5 min in 3× SSC/0.1% SDS at room temperature in a crystallization dish, then wash in the same solution at 65°C for 2 h. ④

3 Pre-hybridize filters in SHB overnight, in a dish covered with plastic film, using gentle agitation at hybridization temperature. ⑤

4 Set up the following reaction mix in a 1.5 ml microcentrifuge tube:
  - 10× kinase buffer   1 µl
  - oligonucleotide   20 pmol
  - [γ -^{32}P]ATP (3000 Ci/mmol,   5 µl (50 µCi, 1.85 MBq)
    111 TBq/mmol)
  - T4 polynucleotide kinase (10 U/µl)   1 µl
  - distilled water x µl (to 10 µl final vol.)

  incubate the mixture at 37°C for 1–2 h, then heat for 10 min at 68°C to inactivate the enzyme. ⑥

5 Purify the radiolabeled oligonucleotide using a Sephadex G-50 spun column (See *Protocol 1*, Note 4). ⑦

6 Hybridize filters for 14–48 h in sealable bags (no more than five to ten filters per bag) containing 20–30 ml of SHB containing 0.02 pmol (for

57

Notes

Time required: 3–5 days depending on the length of hybridization

Day 1 - take replica filter of plaques or bacterial colonies, wash filters, prehybridize overnight.

Day 2 - radiolabel probe, hybridize filter (14–48 h).

Day 3 - wash filters and set up autoradiography.

① The oligonucleotide should be a minimally redundant guessmer at least 32–36 nucleotides in length, designed according to the guidelines of Lathe [5].

② For preparation of TMAC solutions see ref. 6, p.11.50.

③ Prepare filters bearing phage or bacteria according to standard procedures [6]. A step of denaturation should be included to facilitate hybridization of immobilized DNA to the radioactive probe [6].

④ This step removes much of the bacterial debris adhering to the filters.

⑤ Hybridization temperature should be 10–25°C below melting temperature (T_m). T_m can be calculated as follows: $T_m = 102-820/L-1.2 (100-H)$ where L is the length of oligonucleotide, and H is the percentage homology between the probe and the actual sequence, estimated as 85% [5]. This assumes an oligonucleotide with a G+C content of 50%. In practice, for oligonucleotides 32–55 nucleotides in length, a hybridization temperature of 40–55°C is employed.

⑥ Specific activity should be ~3 × 10^6 c.p.m./pmol.

bacterial colonies) to 0.2 pmol (for phage plaque) of radioactive probe per ml (specific activity ~ 3×10^6 c.p.m./pmol).

7 Wash the filters four times for 5 min at room temperature with 6× SSC/0.1% SDS.

8 Wash the filters for 30 min in 6× SSC/0.1% SDS at hybridization temperature. ⑧

9 Check with Geiger counter; if signals are significantly above background, continue washing at 2.5°C increments.

10 Cover the filters with plastic film, and expose overnight to X-ray film with an intensifying screen. ⑨

11 Purify positive phage/bacteria which give signals on duplicate filters, and characterize further.

B. TMAC procedure [18,19]

1 Prepare duplicate nitrocellulose filters (132 mm) bearing bacteriophage plaques or bacterial colonies (up to 10 000/15 cm plate). ⑩ ⑪

2 Wash filters three to five times for 5 min each in 3× SSC/0.1% SDS room temperature, then wash in the same solution at 65°C for 2 h.

3 Submerge each filter and with gloved hand gently rub the filters to remove debris.

⑦ See *Protocol 1*, Note 4.

⑧ Filters should move around freely without sticking to each other.

⑨ Do not permit filters to dry out completely since this leads to irreversible binding of probe.

⑩ Prepare filters bearing phage or bacteria according to standard procedures [6]. A denaturation step should be included to facilitate hybridization of DNA to the radioactive probe [6].

⑪ Nitrocellulose filters become fragile in TMAC solutions, they must be handled carefully. Alternatively, nylon filters can be used, although the protocol is slightly different [20].

⑫ Hybridization temperature is 5–10°C below the irreversible melting temperature (T_i). T_i is calculated as follows: $T_i = 97-682/L$ where L is the oligonucleotide length. This equation was derived experimentally from data for oligonucleotides ranging from 16–32 bases in length, but is probably also valid for longer oligonucleotides [19]. The probe is radiolabeled as described in Procedure A.

⑬ Hybridization in the presence of TMAC is compatible with the use of highly degenerate oligonucleotides, and use of stringent hybridization temperatures (approaching the melting temperature of a perfect hybrid). However, the longer the probe, the higher will be the degeneracy

4 Transfer to THB, 5–10 ml per filter. Solution should be pre-warmed to hybridization temperature. Agitate the filters for 1–2 h. ⑫

5 Transfer the filters to pre-warmed THB hybridization solution containing 1–2 × 10^6 c.p.m./ml labeled oligonucleotide. ⑬

6 Incubate for 48–72 h at hybridization temperature.

7 Wash with TWB several times at room temperature, then for 1 h at 10–15°C below hybridization temperature.

8 Wash three times for 10 min each with 2× SSC/0.2% SDS at room temperature.

9 Purify positive phage/bacteria and characterize further.

and the lower the concentration of correct sequence in the probe. A typical probe will be a 17-mer with up to 512-fold degeneracy.

Protocol 10. Plaque and colony hybridization

Protocol 11. PCR cloning of transcription factors using degenerate primers.

W. M. Macfarlane

Reagents

- Absolute ethanol▽
- Agarose (Sigma)
- 10× Agarose gel loading buffer [0.2 % (w/v) Bromophenol blue, 0.2% (w/v) xylene cyanol, 50% (v/v) glycerol]
- Amplitaq™ polymerase (Cetus)
- cDNA①
- Distilled water
- DNA kilobase ladder markers (Gibco)
- dNTP mix (2 mM each of dATP, dTTP, dCTP, dGTP)
- 70% Ethanol▽
- Ethidium bromide (10 mg/ml)▽
- Mineral oil
- 100 mM Oligonucleotide primer stocks②
- 10× PCR buffer (100 mM Tris-HCl pH 8.8, 500 mM KCl, 30 mM MgCl$_2$, 0.1 % gelatin)
- 3 M Sodium acetate pH 6.0

Equipment

- PCR thermal cycler (Perkin/Cetus)
- Horizontal agarose gel apparatus
- 0.5 ml Microcentrifuge tubes
- Power pack
- Scalpel blades (fine)
- UV light-box▽
- Vacuum flask containing liquid nitrogen▽
- Vortex mixer
- TA vector
- TE buffer (10 mM Tris-HCl pH 8.0, 1 mM EDTA)
- Tris-buffered phenol, pH 8.0▽
- 10× TBE buffer (162 g Tris, 9.3 g EDTA and 26.5 g boric acid – dissolve in 1 liter distilled water, autoclave and store in a plastic bottle)

Procedure

1 To a 0.5 ml microcentrifuge tube, add the following:
- cDNA①   1 μl
- 10× PCR buffer   10 μl
- dNTP mix   10 μl
- oligonucleotide primer A②   1 μl (1 μM final concentration)
- oligonucleotide primer B②   1 μl (1 μM final concentration)
- distilled water   77 μl (to final volume of 100 μl)

overlay with 50 μl of mineral oil, and incubate in a thermal cycling apparatus at 90°C for 10 min.

2 Add 1 U of Amplitaq™ polymerase to each reaction vial.

3 Programe the thermal cycler to repeat the following cycle 35 times:③④
    95°C, 1 min
    40°C, 1 min
    70°C, 30 sec.

4 On completion of this series, programme the cycler to carry out an additional 10 min incubation at 70°C, to polish the DNA ends.

5 Remove 10 μl of each reaction, and mix with 1 μl of 10× agarose gel loading buffer. Analyze on a 1% agarose gel (containing 10 ng/ml ethidium bromide), in 1× TBE buffer, by separating at 70 V for 40 min, with appropriate size markers.

6 Analyze gel on UV lightbox. Cut appropriate bands from the gel using a fine scalpel blade, and dice gel slice into small pieces.

Notes

Time required: 1 day

① cDNA library from tissue of interest.

② A pair of degenerate primers should be approximately the same length, should have a G+C content of about 50%, should not contain long stretches of the same base and should not have significant complementarity to each other, particularly at their 3′ ends. Complementarity here leads to the amplification of 'primer dimers', which are perfect PCR substrates, and will form the dominant product in any reaction.

③ For annealing temperatures, allow 2°C for every A or T, and 4°C for every G or C. For degenerate primers, values must be averaged between the minimum possible annealing temperature (i.e. where the primer would contain the maximum possible number of A–Ts), and the maximum values (i.e. where the primer would contain the maximum possible number of C–Gs) to generate a reasonable starting temperature for the annealing step.

④ If after optimization of PCR conditions a number of seemingly specific bands appear which do not correspond to the desired product, run parallel PCR reactions containing only one primer, as this effect is often caused by specific double priming of one of the primer pairs. Comparison of PCR products of singly and doubly primed reactions allows elimination of many such nonspecific products.

Protocol 11. PCR cloning of transcription factors

7 Add 500 µl of Tris-buffered phenol to each sample, and vortex hard for 1 min. Immerse samples in liquid nitrogen for 15 min, then centrifuge at 10 000 g for 30 min in a microcentrifuge at 4°C.

8 Ethanol precipitate PCR products from the aqueous phase. Add 2 vol. ethanol, 0.1 vol. 3 M sodium acetate, and incubate at -70°C for 1 h. Centrifuge at 10 000 g for 30 min. Wash pellet with 500 µl of 70 % ethanol, air dry.

9 Resuspend pellet in 10 µl of TE buffer, and ligate into TA vector for sequencing. ⑤

⑤ Following appropriate enrichment procedures, such as length selection by agarose gel electrophoresis, PCR fragments can be blotted and hybridized by a gene specific probe, or more commonly, subcloned into a suitable vector for sequencing. Cloning vectors are now commercially available, which take advantage of the 3′ overhanging As which are the natural result of PCR amplification with *Taq* polymerase (such as Invitrogen's TA Cloning Vector), and these T-vectors [21] allow direct cloning of PCR products.

Reagents

Bacteriophage expression cDNA library (λgt11 or equivalent)
Blotto [50 mM Tris-HCl pH 7.5, 1 mM EDTA, 1 mM DTT, 5% (w/v) milk powder]
Escherichia coli Y1090 or equivalent
Guanidine hydrochloride (GHCl)(1)
Hepes binding buffer (HBB) (25 mM Hepes pH 7.9, 25 mM NaCl, 5 mM $MgCl_2$, 0.5 mM DTT)
Instant nonfat milk powder
Isopropyl-β-D-thiogalactopyranoside (IPTG) (10 mM)
Radiolabeled concatenated oligonucleotide probe(2)
Sonicated, denatured salmon sperm DNA (SS DNA)

Tris-binding buffer (TBB) (10 mM Tris-HCl pH 7.5, 1 mM EDTA, 1 mM DTT)

Equipment

Crystallization dishes or petri dishes diameter 150 mm
Incubators (42°C and 37°C)
Nitrocellulose filters 0.45 mm, 132 µm diameter
Orbital platform shaker
Plastic film
Waterproof ink
Whatman 3MM filter paper

Procedure

A. Using denaturing conditions [12]

1 Infect host strain (*E. coli* Y1090) with bacteriophage library.

2 Apply to 15 cm plates at 25 000 plaque-forming units per plate.

3 Incubate at 42°C for 4 h.

4 Wet a nitrocellulose filter (132 mm diameter) in 10 mM IPTG solution, and remove excess liquid by placing on 3MM paper.

Notes

Time required: 3 days

(1) Ultra-pure grade guanidine hydrochloride is generally recommended. However, less expensive grades (e.g. US Biochemicals 'Practical' grade) have been used successfully.

(2) ^{32}P-labeled probe is prepared by nick translation [12] or end-labeling [13] of concatenated binding sites (see *Protocol 9*).

(3) To prevent lifting of top agar with the filter, plates may be

5 Place the filter carefully on the plate.

6 Incubate at 37°C for 6 h.

7 Orient the filter on the plate by stabbing through the filter at three locations around the periphery with a syringe needle containing waterproof ink.

8 Lift the first filter from the plate and air dry on 3MM paper with plaque side up. ③

9 Apply a new IPTG-impregnated filter to the plate.

10 Mark the second filter with a needle at the same locations as the initial marks.

11 Incubate at 37°C for 2 h.

12 Lift the second filter and air dry it on 3MM filter for at least 15 min. ☐1

13 Put filters in a dish with HBB containing 6 M GHCl.

14 Agitate the filters gently on an orbital shaker for 5 min at 4°C to remove bacterial debris.

15 Replace with fresh HBB containing 6 M GHCl, and agitate gently for 5 min at 4°C.

16 Pour the solution into a measuring cylinder, dilute 1:1 with HBB, mix, pour back into the dish, and agitate the filter gently for 5 min at 4°C.

cooled for 10 min at 4°C just before lifting.

④ Filters should move around freely without sticking to each other.

⑤ True positives typically give a 'doughnut' shaped autoradiographic image whereas spots with dark centers are often false positives.

⑥ Filters can be stored in TBB overnight at 4°C prior to incubation with probe.

⑦ Filters should move around freely without sticking to each other.

⑧ True positives typically give a 'doughnut' shaped autoradiographic image whereas spots with dark centers are often false positives.

17 Repeat Step 16 four times.

18 Replace the solution with HBB to rinse out traces of GHCl and gently agitate the filter for 5 min at 4°C.

19 Repeat Step 18.

20 Transfer the filters to a new dish with HBB containing 5% (w/v) milk powder.

21 Agitate the filters gently for 30 min at 4°C.

22 Replace the solution with HBB containing 0.25% (w/v) milk powder, and agitate the filters gently for 5 min at 4°C.

23 Repeat Step 22.

24 Transfer the filters to a new dish containing ^{32}P-labeled, concatenated DNA probe (10^6 c.p.m./ml) in HBB containing 0.25% (w/v) milk powder.

25 Agitate the filters gently for 2–12 h at 4°C.(4)

26 Wash the filters at room temperature for 5–10 min with HBB containing 0.25% (w/v) milk powder (the solution is kept on ice).(4)

27 Repeat Step 26 four times.

28 Place the filters on 3MM paper to remove excess liquid.

29 Cover with plastic film, and expose to X-ray film overnight with an intensifying screen.(5)

30 Plaque purify and characterize clones.

Pause point

[1] Filters can be stored dry overnight at 4°C prior to incubation with probe.

B. Using nondenaturing conditions [13]

1 Follow Steps 1–12 as in Procedure A, except that filters are not allowed to dry. After lifting from plates immerse the filters directly in a dish containing Blotto.

2 Incubate for 60 min at room temperature with gentle agitation on an orbital shaker.

3 Transfer to a dish containing TBB and agitate the filters gently for 5 min. ⑥

4 Repeat twice with fresh TBB.

5 Transfer the filters to a new dish containing ^{32}P-labeled, concatenated DNA probe (10^6 c.p.m./ml) in TBB containing 5 μg/ml SS DNA.

6 Agitate the filters gently at room temperature for 60 min. ⑦

7 Wash filters four times for 7.5 min each in TBB. ⑦

8 Cover with plastic film, and expose to X-ray film with an intensifying screen overnight. ⑧

9 Purify plaques that produce a signal on duplicate filters, and characterize further.

IV IDENTIFICATION OF TRANSCRIPTION FACTORS BY DATABASE HOMOLOGY SEARCHING.   P.A. Moore

Introduction

The recent increase in the volume of genetic information generated from prokaryotic and eukaryotic genomic and cDNA sequencing projects has enabled scientists to identify many novel genes and their possible function, based on their sequence homology to known genes. This approach followed by subsequent functional studies, has led to the identification of several novel transcription factors including the isolation of the human homologs of the yeast transcription factors TOA1 [1], ADA2 [2], GCN5 [2] and the isolation of IRF3, a novel interferon regulatory factor [3]. As the volume of genetic information increases from various sequencing projects, and the ease with which this sequence information can be manipulated improves, this approach towards the identification of transcription factor homologs will become increasingly useful. The powerful and user-friendly resources provided by the World Wide Web allow this type of search to be performed via a multitude of overlapping avenues and one approach is described here.

Protocol provided

13. *Identification of transcription factors by database homology searching*

References

1. Ozer, J., Moore, P.A., Bolden, A.H., Lee, A., Rosen, C.A. and Lieberman, P.M. (1994) *Genes Dev*. **8**:2324.
2. Candau, R., Moore, P.A., Wang, L., Barlev, N., Ying, C.Y. and Rosen, C.A. (1996) *Mol. Cell. Biol*. **16**:593.
3. Au, W.-C., Moore, P.A., Lowther, W. and Pitha, P.M. (1995) *Proc. Natl Acad. Sci. USA*, **92**:11657.

Protocol 13. Identification of transcription factors by database homology searching. P.A. Moore

Equipment

Computer linked by Netscape (or equivalent browser) to World Wide Web

Procedure

1 Retrieve the nucleotide sequence of the transcription factor for which you intend to obtain a homolog. Sequences, if not already on computer format, can be obtained from a variety of sources including the Genodatabase site on the World Wide Web (http://specter.dcrt.nih.gov:8004/) and should be saved on to computer hard disk or on to a floppy disk. ①

2 Open the saved sequence file as a Microsoft Word document (or equivalent) and edit the nucleotide sequence to remove any numbers or spaces between nucleotides.

3 Copy the edited sequence from the Word document and paste it into the space provided within the NCBI-BLAST Notebook page located on the World Wide Web (http://www.ncbi.nlm.nih.gov/Recipon/bs_seq. html). ②

4 Select the BLAST algorithim and the sequence database you wish to use to perform the homology search, set the parameters controlling the BLAST search and the results output, and then execute the homology search by

Notes

① The Genodatabase site provides access to a range of databases including those provided by GenBank (the US National Institutes of Health genetic sequence database, maintained by the US National Center for Biotechnology Information, NCBI; http://www.ncbi.nlm.nih.gov/) and EMBL (the European Molecular Biology Laboratory in Heidelberg, Germany; http://www.embl-heidelberg.de/). DNA or amino acid sequences are retrieved most easily by entering the accession number, but can also be accessed by using an author's name or a description of the sequence.

② The Basic Local Alignment Search Tool (BLAST) provides a reliable algorithm for searching databases relatively quickly for sequence homology. More specialized algorithms for performing homology searches are available, but are significantly slower than BLAST and rarely reveal significant relationships missed by BLAST. The time taken for the search to be completed depends on the length of the query sequence, the complexity of the search (dictated by the parameters) and how busy the server is at the time of the search.

hitting the 'Perform Search' icon. While the default parameters (defined by Hypertext links to the NCBI-BLAST notebook) which regulate the BLAST search and the results output are sufficient for most searches, these parameters can be refined if necessary.③④

5 The results of the BLAST analysis are displayed on the subsequent Web page and should be examined carefully to determine if a novel gene homolog has been identified. Homology between the query sequence and the identified sequence should exist both at the nucleotide level and protein level and should be maintained throughout the coding regions of both genes. If the homology holds up, then the novel gene should be obtained, if possible from the source which generated the sequence information, and subjected to functional studies. In cases where the homolog is an EST, it is often necessary to first isolate a full length cDNA clone before pursuing biological characterization.

③ Various versions of BLAST are available: BLASTN compares a nucleotide query sequence against a nucleotide database(s); BLASTX compares all six open reading frames (ORFs) of a nucleotide query sequence against a protein database(s); TBLASTX compares all six ORFs of a nucleotide query sequence against all six ORFs of a nucleotide database(s); BLASTP compares a protein query sequence against a protein database(s); and TBLASTN compares a protein query sequence against all six open reading frames of a nucleotide database.

④ The NCBI-BLAST notebook allows you to search a wide range of databases. Of particular use for the identification of novel transcription factors are EST (expressed sequence tag) databases, which are generated by high throughput sequencing of random cDNA clones, and which can be screened against using the NCBI-BLAST notebook. In some instances, it is desirable to search a specific organism for a gene homolog. Several organisms' genomes are close to, or have been completely sequenced, and more specialized servers are available to perform BLAST searches on these databases. BLAST servers are available (URL in parentheses) for sequence databases created from *Caenorhabditis elegans* (http://www.sanger.ac.uk/~sjj/C.elegans_blast_server. html); *Haemophilus influenzae* (http://www.tigr.org/ tdb/mdb/hidb/hidb/_seq_search/hidb_seq_search/.html); *Schizosaccharomyces pombe* (http://www.sanger.ac. uk/yeast/S.pombe_blast_server.html); *Mycobacterium tuberculosis* (http://www.sanger.ac.uk/pathogens/ TB_blast_server.html); and *Saccharamoyces cerevisiae* (http://genome-www.stanford.edu/Saccharomyces/).

Protocol 13. Identification of transcription factors

V MAPPING REGULATORY SEQUENCES. S. Goodbourn

Introduction

One of the most fundamental approaches to the study of gene regulation is the delineation of those sequences that govern the tissue- or ligand-specific expression of the gene of interest. These sequences are referred to as *cis* acting, and experience has shown that these can be classified as promoter elements, which lie adjacent to the mRNA startpoint, and enhancer elements, which are often remote from the transcription unit. The standard procedure for defining the *cis*-acting sequences is to first identify a fragment of DNA that can be introduced into cell culture and still reproduce the regulatory property of interest, and then to pare down this DNA fragment until it no longer functions. Obviously the simplest way to do this is to make progressively smaller restriction fragments. However, regulatory elements are often rather small (such elements reflect the binding of DNA sequence-specific transcription factors, or *trans*-acting factors, which usually recognize an 8–12 bp sequence) and often restriction enzyme sites are not conveniently located to allow a sufficiently detailed resolution. To get around this limitation, a number of enzymatic procedures have been developed that allow the controlled removal of sequences from the 5′ ends of DNA molecules to generate deletions. The effect of these deletions are then analyzed in transfection assays. This is shown schematically in *Figure 2*, where the function of a given test gene is observed in an initial experiment with a construct containing 500 bp of sequence to the 5′ side of the mRNA startpoint, or cap site (indicated as −500 in construct A). Function is retained when the 5′ flanking region is shortened to −300 (construct B), but lost when the region is shortened to −200 (construct C). It is important to stress that the loss of function observed between constructs B and C does not necessarily imply that the key regulatory sequences lie between −300 and −200, but merely that function has been inactivated by a deletion to −200, it is possible that the −200 region actually spans the regulatory element.

While these assays determine whether a sequence is necessary for regulation, they do not establish whether it is sufficient, and this is generally achieved by fusing the sequence of interest to a reporter gene (a description of reporter genes and

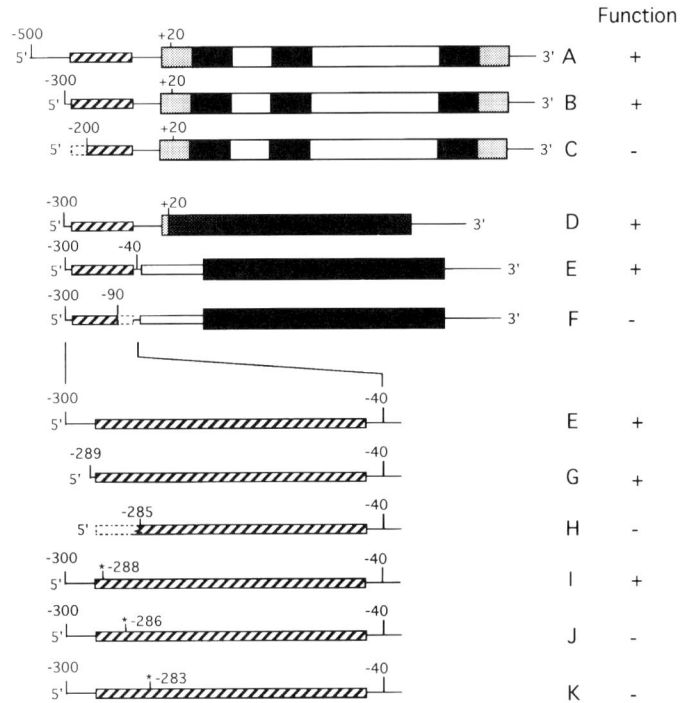

Figure 2. Structure–function of regulatory regions.

methods used to transfect mammalian cells in culture is provided in the companion volume *Gene Transcription: RNA Analysis*). This is shown in *Figure 2,* construct D, in which a -300 to $+20$ fragment can confer the appropriate regulatory properties on a heterologous gene. In many cases, the reporter gene has its own basic signals for the specification of transcription [e.g. the thymidine kinase gene of herpes simplex virus (HSV tk) is often used] with the result that it is possible to cut down considerably on the amount of DNA required for regulation, and the effects of deletions from the 3′ end of the regulatory sequence can be determined. Thus in *Figure 2* construct E contains a -300 to -40 fragment which can confer regulation upon a heterologous promoter. It is then possible to perform a 3′ deletion series on the $-300/-40$ region to define the maximal regulatory element, for example, a 3′ deletion to -90 inactivates function (*Figure 2,* construct F); taken together with the 5′ deletion study, this indicates that all the germane regulatory sequences reside between -300 and -40 (see striped box in *Figure 2*).

The above procedure can be regarded as an essential first step in defining a regulatory element. In many cases, a greater level of detail will be required before proceeding with an analysis of *trans*-acting factors. This 'second phase' analysis is similar to the first, except that more closely clustered deletion breakpoints are required, and it is often advantageous to introduce point mutants, either at defined sites or at random, into the region of interest. Continuing the analysis of a fictitious promoter, *Figure 2,* construct G can still be seen to function, whereas construct H does not. During the first phase of analysis we could only map the 5′ end of the regulatory element to between -300 and -200, however, this second phase analysis allows us to position it to between -289 and -285. A single base change (point mutation) at -288 has no effect on regulation (*Figure 2,* construct I), while changes at -286 and -283 (*Figure 2,* constructs J and K) abolish regulation. These mutations allow us to confirm that the regulatory element actually spans the -285 deletion breakpoint of construct H, rather than being contained between -289 and -285.

An alternative to single base substitutions is the analysis of clustered point mutations. These are best achieved using the so-called 'linker scanning' mutations. In this case, deletions with linkers at defined 5′ and 3′ ends are recombined so that a region of the regulatory element is neatly replaced with a linker sequence (see *Figure 3*). Alternatively, mutants can be

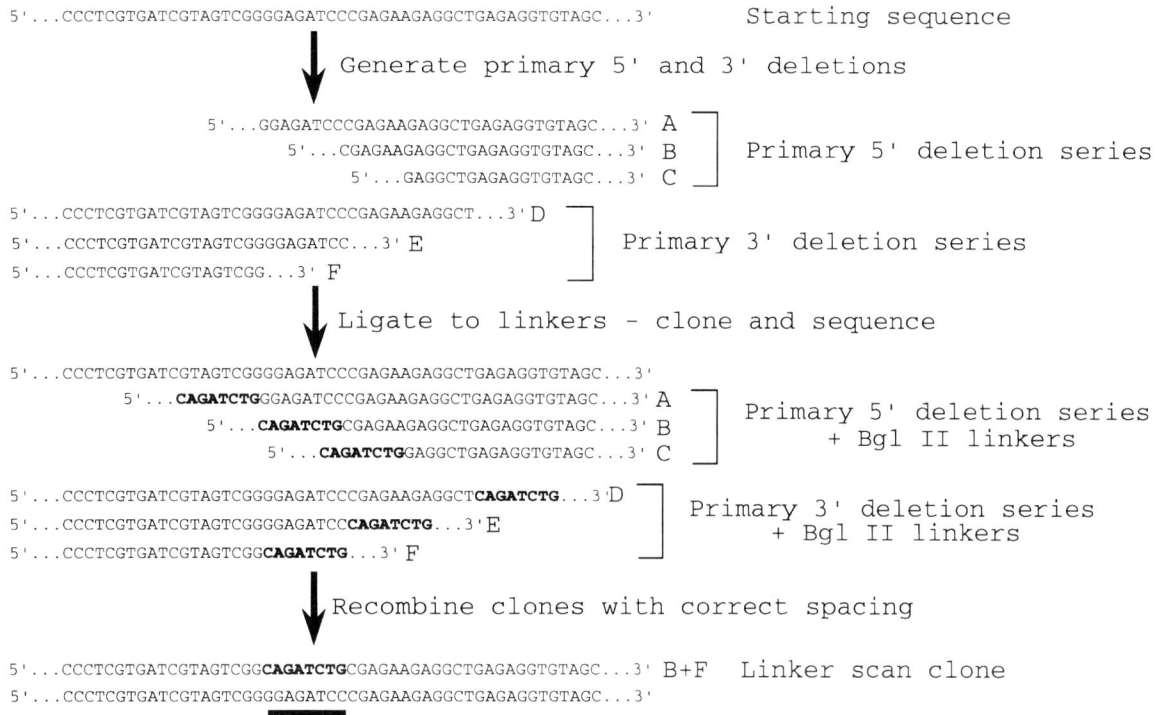

Figure 3. Construction of linker-scanning mutants.

Mapping regulatory sequences

recombined to generate internal deletions or duplications. This approach has been optimally used by McKnight [1] to define the HSV tk promoter. A potential drawback of the procedure is that the construction of a linker scan series which conserves the original spacing of the regulatory element requires deletion mutants with precise endpoints. This can be rather difficult to achieve owing to the sequence-specific effects of the various nucleases (see below).

The various methods for introducing deletions into a DNA target are discussed in the next section, and protocols for nuclease Bal 31 digestion (*Protocol 14*) and targeted deletion mutagenesis (*Protocol 15*) are presented. A PCR-based method for the introduction of defined point mutations is presented in *Protocol 16*, and a method for random mutagenesis is presented in *Protocol 17*.

Deletion mutagenesis

By far the most widely applied method for generating 5′ deletions is the use of nuclease Bal 31 (ref. 2; see *Protocol 14*). This enzyme contains a double-stranded DNA-specific 3′ exonuclease activity, and a single-stranded DNA-specific 3′ endonuclease activity, which together allow double-stranded DNA to be degraded from each end [3]. An alternative approach is to use exonuclease III, which is a single-stranded exonuclease that removes DNA from the 3′ end of double-stranded DNA; this enzyme must be used in conjunction with an additional enzyme (Mung bean nuclease) to remove the 5′ overhanging DNA. A similar approach to the use of exonuclease III is to use the 3′ exonuclease activity of DNA polymerases such as T4 DNA polymerase, and of course these enzymes must also be used in conjunction with Mung bean nuclease.

Any of these approaches can be used to generate a relatively crude deletion series in which the endpoints are separated by tens to hundreds of base pairs (suitable for the 'phase one' analysis described above). However, in each case the enzyme reactions are difficult to control accurately, and more importantly the nucleases show distinct sequence specificity. Thus, Bal 31 tends to generate deletions with G or C at the 5′ end while exonuclease III and T4 DNA polymerase generates deletions with 5′ ends which map to A residues; since the latter enzymes are 3′ exonucleases, these data suggest that they are inefficient at removing T residues [4]. These limitations are sufficient to prevent the generation of a closely nested deletion series in any given region of interest, and also present a serious obstacle to the production of accurate linker scan mutants. A method to circumvent this is presented in *Protocol 15*. In this procedure (outlined in *Figure 4*), the region of interest is cloned into a vector that allows the production of single-stranded DNA (e.g. the M13 vectors). The single-stranded DNA is then hybridized to an oligonucleotide that binds to sequences adjacent to the target region (this is easily achieved by cloning the region of interest into a standard M13 vector such as M13mp19 [5] and priming with the universal sequencing primer), and the primer is extended using Sequenase T7 DNA polymerase. During this polymerase reaction, one of the four dNTPs is contaminated with a small amount of the equivalent α-thio-dNTP so that, on average, one such nucleotide is incorporated into the target region per molecule. The template is then cut with a restriction enzyme that delineates the 5'-most point of the deletion series, and the templates are digested with exonuclease III. The key to this procedure is that exonuclease III is unable to degrade DNA wherever α-thio-dNTP has been incorporated, which results in the generation of base-specific 3' deletions. The residual 5' overhangs are removed by Mung bean nuclease, and the products cloned with the desired linkers. Using this approach it has proved possible to generate deletions with most 5' endpoints over a 40 bp target [4]. It has also significantly eased the production of linker scan mutants (unpublished observations).

Single base substitutions

Methods for the substitution of single nucleotides within a defined region fall into two categories; mutations are either introduced at random, or they are introduced at specific sites. Random mutagenesis can be achieved *in vivo*, although the efficiencies of known mutagenic processes are generally not high enough to saturate a specific target with mutations. The alternative is to generate mutations *in vitro*. DNA can be damaged by a number of agents, and this damage is often not repaired correctly in bacteria, leading to the fixing of mutants. Early methodology for random mutagenesis concentrated on the use of double-stranded DNA as the target for mutagenesis. However, this approach suffers from the fact that if only one strand is damaged then repair is likely to restore the original sequence. Furthermore, since most mutagens modify the purine or pyrimidine rings, these are generally not freely accessible in nonreplicating DNA. The use of single-stranded DNA tem-

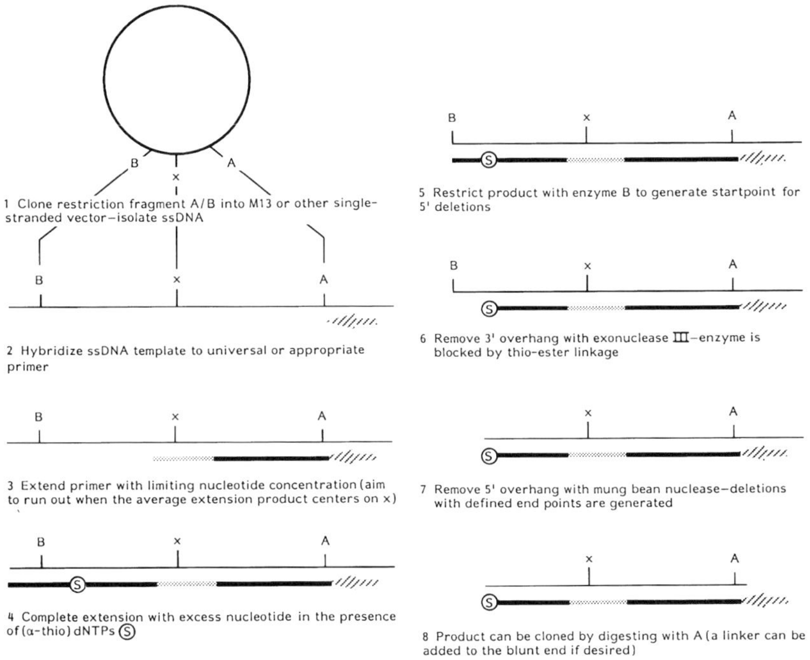

Figure 4. Construction of sequence-specific deletion mutants.

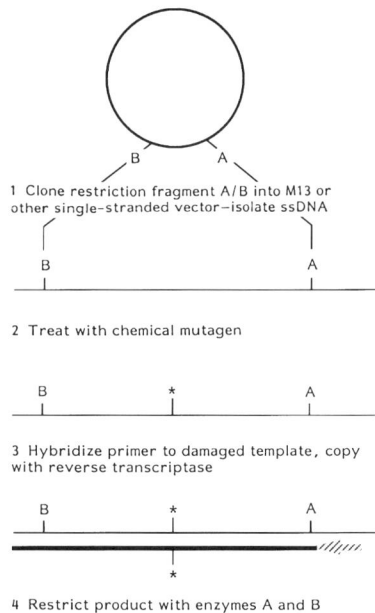

1 Clone restriction fragment A/B into M13 or
other single-stranded vector—isolate ssDNA

2 Treat with chemical mutagen

3 Hybridize primer to damaged template, copy
with reverse transcriptase

4 Restrict product with enzymes A and B

Figure 5. Generation of random single box substitutions in a defined genetic target.

plates much improves the efficiency of mutagenesis. An effective method [6] for random mutagenesis of a defined target is presented in *Protocol 16* and is illustrated in *Figure 5*. The target region (defined by restriction enzyme sites A and B in *Figure 5*) is first cloned into an M13 vector or equivalent, and a large-scale preparation of single-stranded DNA is made. The single-stranded DNA is then treated with chemical mutagens, which cause base-specific modification. This is exemplified by formic acid which depurinates DNA without leading to breakage of the sugar–phosphate backbone. The damaged template is then copied from a fixed primer using reverse transcriptase which is capable of copying depurinated and modified templates. The reverse transcriptase incorporates a nucleotide at random into the gap left by the damaged base. Finally, the double-stranded DNA fragment is removed from the M13 template by restriction digestion with enzymes A and B and is recloned into a suitable vector backbone. The extent of the mutagenesis can be varied according to the target size, although any calculation should balance the conflicting requirements that a significant portion of the clones should contain mutations that have the possibility of multiple mutations. An excellent resolution of these difficulties can be achieved by coupling the mutagenesis method with a method which allows the preparative purification of mutant DNA, such as preparative denaturation gradient gel electrophoresis [6], although these approaches can be technically demanding.

Targeted mutagenesis using PCR. M.C.U. Hammond-Kosack

In many applications, it is not necessary to generate a library of point mutations, and many methods have been devised for targeted mutagenesis. Until recently, the most effective of these was based upon primed synthesis directed from a single-stranded DNA template using primers containing defined base changes. However, the advent of PCR has led to huge improvements in efficiency for targeted mutagenesis, and a method for this is shown in *Figure 6*.

Four primers are needed for the synthesis of one mutation in a promoter fragment. Primers R1 and R2 flank the region and contain suitable restriction enzyme sites for cloning after PCR. Primer M consists of three parts: the 5′ and 3′-most parts are identical to the wild-type promoter sequences flanking the sequence to be altered, and the middle part contains the mutated nucleotide sequence. Primer rev is complementary to the wild-type 5′ part of primer M which serves as an anchor for the regeneration of the full length promoter fragment.

The protocol consists of three steps, all of which employ *Taq* DNA polymerase. The first step consists of two separate PCR reactions, using the primer pairs M–R2 and R1–rev and plasmid DNA containing the wild-type promoter sequence as a template. This generates two PCR fragments, A and B, which represent the upstream wild-type part (fragment A) and the mutated downstream part (fragment B) of the promoter. Both fragments have a common sequence, the anchor, at their 3′ and 5′ end, respectively. After gel purification of these two DNA fragments, to remove the residual primers and the wild type plasmid template, the second step consists of annealing the two fragments and extending the annealed heterodimer with *Taq* DNA polymerase without the addition of any primers. Fully extended heterodimers should contain the mutated nucleotides as well as the restriction sites R1 and R2 on either end. During the third step the primers R1 and R2 are added and the mutated promoter sequence is then amplified by PCR. The resulting PCR fragment is digested with the restriction endonucleases which cleave the terminal sites created by primers R1 and R2, gel purified and cloned into a vector [7–9].

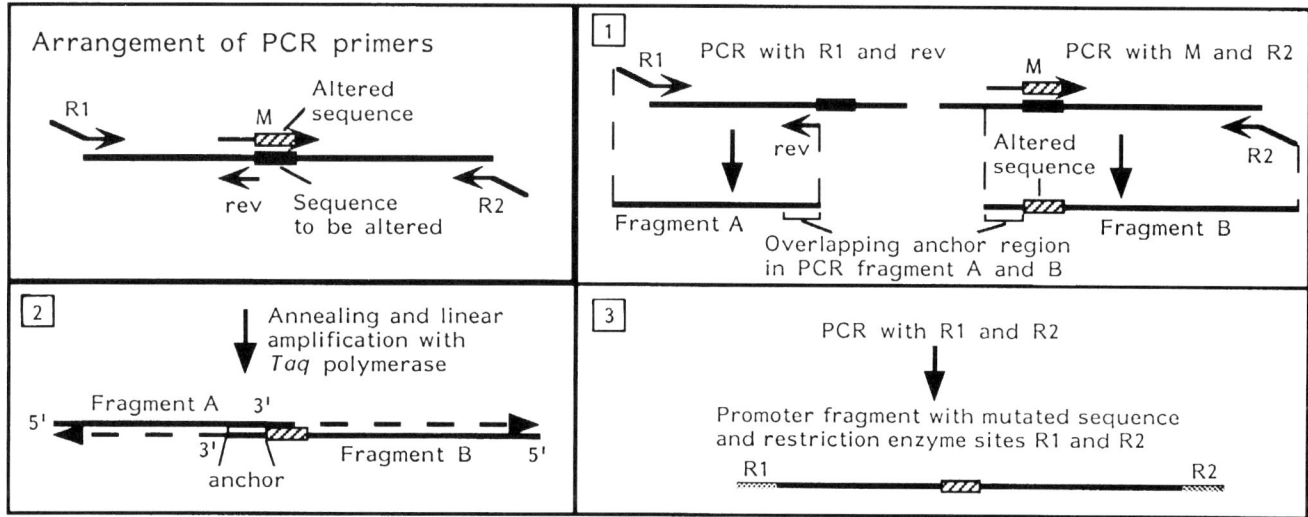

Figure 6. Arrangement of PCR primers for targeted mutagenesis, see text for further details.

Mapping regulatory sequences

Protocols provided

14. *Generation of deletions using Bal 31*
15. *Generation of sequence-specific deletion mutants*
16. *Single-stranded DNA mutagenesis*
17. *Targeted mutagenesis using PCR*

References

1. McKnight, S.L. and Kingsbury, R. (1982) *Science,* **217**:1193.
2. Legerski, R.J., Hodnett, J.L. and Gray, H.B., Jr (1978) *Nucleic Acids Res.* **5**:1445.
3. Lau, P.P. and Gray, H.B., Jr (1979) *Nucleic Acids Res.* **6**:331.
4. King, P. and Goodbourn, S. (1992) *Nucleic Acids Res.* **20**:1039.
5. Yanisch-Perron, C., Viera, J. and Messing, J. (1985) *Gene,* **33**:103.
6. Myers, R.M., Lerman, L.S. and Maniatis, T. (1985) *Science,* **229**:242.
7. Higuchi, R., Krummel, B. and Saiki, R.K. (1988) *Nucleic Acids Res.* **16**:7351.
8. Rashtchian, A., Thornton, C.G. and Heidecker, G. (1992) *PCR Methods Appl.* **2**:124.
9. Valette, F., Mege, E., Reiss, A. and Adesnik, M. (1989) *Nucleic Acids Res.* **17**:723.

Mapping regulatory sequences

Protocol 14. **Generation of deletions using Bal 31.** S. Goodbourn

Reagents

Absolute ethanol⚠

$5\times$ Bal 31 reaction buffer (1 M NaCl, 60 mM $CaCl_2$, 60 mM $MgCl_2$, 100 mM Tris-HCl, 5 mM EDTA, pH 8.0)

500 mM EGTA, pH 8.0

70% Ethanol⚠

Linearized plasmid DNA molecule (1 μg/ml stock)

Nuclease Bal 31

Phosphorylated linkers (e.g. Bgl II linkers at 200 ng/μl)

Restriction enzymes and buffers

Sterile deionized water

Sterile TE (10 mM Tris-HCl, pH 7.4, 1 mM EDTA)

T4 DNA ligase and ligase buffer

Tris-equilibrated phenol⚠

Yeast tRNA (10 mg/ml in TE)

Equipment

Ice-water bath

Set of micropipettes (e.g. Gilsons or Finn pipettes)

Microcentrifuge

1.5 ml Microcentrifuge tubes

Vortex apparatus

Water bath set at 30°C

Yellow and blue polypropylene tips for micropipettes

Procedure

1 To a 1.5 ml microcentrifuge tube, add 10 μg of template DNA that has been linearized at the required startpoint for deletion. Add

- $5\times$ Bal 31 reaction buffer   40 μl
- deionized water   to a volume of 200 μl
- nuclease Bal 31   0.5 U

incubate at 30°C.①

Notes

Time required: 3 days

① The conditions described in this procedure have been optimized for a 4 kb plasmid, and Bal 31 should remove DNA from each end of the linearized plasmid at a rate of 5 bp per min. If the plasmid is smaller than 4 kb, then under these conditions the deletion rate will be slower in

2 Withdraw 40 μl aliquots of reaction mixture at fixed times, ② and terminate the reaction by adding 60 μl of TE + 4 μl of 500 mM EGTA, which specifically chelates the essential cofactor Ca^{2+}. Transfer reactions to ice.

3 When all the required time points have been completed, stop reactions by phenol extraction with 2 vols of buffered phenol, followed by ethanol precipitation in the presence of 20 μg tRNA carrier. 1 Wash pellets in 70% ethanol.

4 Redissolve pellets in 100 μl of deionized water. 1 Examine the extent of deletion by cutting 50 μl of the reaction mixture with a second restriction enzyme that cuts to the 3′ side of the first cut. The products will form a nested set of fragments that vary in size according to the extent of Bal 31 digestion. Depending on the extent of deletion these products can be visualized on either agarose or polyacrylamide gels.

5 Identify the aliquots that have given the desired range, or if this has not been obtained, repeat the nuclease Bal 31 digestion with different extents of incubation. Ligate the remaining 50 μl of reaction mixture to 20 ng of phosphorylated linkers. ③ The mixture can either be transformed directly into *Escherichia coli* (in which case the characterized deletions would have to be recloned into the final vector, since Bal 31 deletions are bi-directional) or the ligated products can be cut with the linker restriction enzyme, plus a second enzyme, and purified before being cloned into the final vector backbone. The former procedure is more efficient, but involves more downstream subcloning.

rough proportion to the size of the plasmid; if the plasmid is larger, the deletion rate will be proportionally faster.

② Times will vary depending upon the required degree of deletion (see Note 1). Thus with a 4 kb plasmid, if deletions of 100–200 bp are required, time points in the range of 20–40 min would be optimal.

③ A titration of linker concentration is recommended.

Pause point

1 May be stored indefinitely at $-20°C$.

Protocol 14. Generation of deletions using Bal 31

Protocol 15. **Generation of sequence-specific deletion mutants.** S. Goodbourn

Reagents

Absolute ethanol⚠
Brain–heart infusion broth
Chloroform⚠
α-^{32}P-dATP (3000 Ci/mmol 111 TBq/mmol, 10 mCi/ml, 370 MBq/ml)⚠
200 nM dCTP
200 nM dGTP
200 nM dTTP
100 mM Dithiothreitol (DTT)
70% Ethanol⚠
10× Exonuclease III reaction buffer (500 mM Tris-HCl, 50 mM MgCl$_2$, pH 8.0)
Exonuclease III (Stratagene)
F positive *E. coli* strain (e.g. JM101)
10× Hybridization buffer (400 mM Tris-HCl pH 7.5, 200 mM MgCl$_2$ and 500 mM NaCl)
L broth (LB)
LB agar
5 M LiCl
10× Mung bean nuclease buffer (300 mM sodium acetate, 500 mM NaCl, 10 mM ZnCl$_2$, pH 4.6)

3 M Sodium acetate, pH 5.2
10 mM Solutions of each α-thio-dNTP (Pharmacia)
Sterile deionized water
TE (10 mM Tris-HCl, pH 7.4, 1 mM EDTA)
α-Thio-dGTP/dNTP mix (20 μM α-thio-dGTP, 60 μM dGTP, 80 μM dATP, 80 μM dTTP, 80 μM dCTP in 50 mM NaCl)
α-Thio-dATP/dNTP mix (20 μM α-thio-dATP, 60 μM dATP, 80 μM dGTP, 80 μM dTTP, 80 μM dCTP in 50 mM NaCl)
α-Thio-dTTP/dNTP mix (20 μM α-thio-dTTP, 60 μM dTTP, 80 μM dATP, 80 μM dGTP, 80 μM dCTP in 50 mM NaCl)
α-Thio-dCTP/dNTP mix (20 μM α-thio-dCTP, 60 μM dCTP, 80 μM dATP, 80 μM dTTP, 80 μM dGTP in 50 mM NaCl)
Tris-equilibrated phenol
1 M Tris-HCl (pH 9.0)
Yeast tRNA (10 μg/μl in TE)

Equipment

9 cm Bacterial petri dishes
Elutips (Schleicher and Schuell)
Ice-water bath
37°C Incubator
Microcentrifuge

Mung bean nuclease (FPLC-purified, Pharmacia)
NaCl (solid)
PEG-8000 (solid)
Phage-specific primer
Phosphorylated linkers (e.g. Bgl II linkers at 200 ng/μl)
Recombinant M13 phage containing fragment to be mutagenized
20% SDS
Sequenase (Amersham/USB Corporation)
SM (5.8 g NaCl, 2.0 g MgSO$_4$.7H$_2$O, 0.1 g gelatin per liter in 50 mM Tris-HCl, pH 7.5)

1.5 ml Microcentrifuge tubes
Perspex radiation shield
Preparative centrifuge (e.g. Beckman J6) and appropriate tubes
15 ml Snap-cap tubes
Shaking incubator and conical flasks (500 ml and 2 liter)
Spectrophotometer and cuvettes
Set of micropipettes (e.g. Gilsons or Finn pipettes)
Vortex apparatus
Water baths set at 16°C, 37°C, 65°C
Yellow and blue polypropylene tips for micropipettes

Procedure

Large-scale preparation of ssDNA

1 Plate out the recombinant phage containing the insert to be mutagenized on a lawn of host *E. coli* (e.g. JM101), and leave at 37°C overnight. Pick a single, well isolated plaque, and in a 15 ml snap-cap tube, grow in 3 ml of brain–heart infusion broth.

2 Whilst the recombinant phage culture is growing (see Step 1) inoculate 100 ml of brain–heart infusion broth in a 500 ml conical flask with a single colony of the host strain of *E. coli* (e.g. JM101). Grow with aeration at 37°C until the culture reaches an OD$_{600}$ of 0.2. Inoculate this with 1 ml of the saturated phage culture and continue to grow at 37°C for 30 min.

3 Transfer the culture from Step 2 to 1 liter of brain–heart infusion broth in a 2 liter conical flask and grow overnight with aeration at 37°C.

Notes

Time required: 7 days

① Under these conditions, the DNA polymerase will extend the primer by 20–80 nucleotides before the supply of dNTPs is exhausted. The templates are thus 'labeled' for analytical purposes without the products extending into the target region.

② Under these conditions, a specific α-thio-dNTP is incorporated into about 1/10–1/20 of the positions normally occupied by the equivalent dNTP. Thus for a target with 50% G–C content, one α-thio-dNTP residue is incorporated every 40–80 bases. Since exonuclease III is inefficient at removing α-thio-dNTPs from DNA, it will only degrade back to the α-thio-dNTP residue closest to the restriction cut generated by enzyme B, and thus the

Protocol 15. Generation of sequence-specific deletion mutants

4 Precipitate bacteria by centrifugation for I0 min at no less than 10 000 *g*. Discard pellet.

5 To the supernatant, add PEG-8000 and NaCl to final concentrations of 4% (w/v) and 0.5 M respectively. Allow to stand overnight at 4°C. [1]

6 Precipitate phage by centrifugation for 20 min at no less than 10 000 *g*. Carefully remove supernatant. Resuspend pellet in 20 ml of SM. [2]

7 Extract with an equal volume of chloroform, and then extract with an equal volume of phenol.

8 Add 2 ml of 3 M sodium acetate (pH 5.2) then 50 ml of absolute ethanol. Store at −20°C for at least 1 h. Recover single-stranded DNA by centrifugation at no less than 10 000 *g*. Wash pellet with 70% ethanol, and dissolve in 5 ml TE. Quantitate DNA recovery using an OD_{260} of 1.0 as equivalent to 36 μg/ml. [3]

Thio-dNTP substitution of single-stranded DNA

1 In a 1.5 ml microcentrifuge tube anneal 1 μg of single-stranded DNA template to 5 ng of an 18 base primer by allowing the primer and template to cool slowly from 65°C to room temperature in 1× hybridization buffer in a 10 μl reaction volume.

2 Extend primer for 5 min at room temperature by adding:
- DTT   1.0 μl of 100 mM
- α-^{32}P-dATP   0.5 μl
- dGTP   0.5 μl of 200 nM

deletion series that is generated from these reaction conditions will only extend up to 80 bp. If a different range of deletions is required, the ratio of α-thio-dNTP to dNTP can be altered so that α-thio-dNTPs are incorporated into the template more frequently (smaller deletions) or less frequently (larger deletions).

③ Avoid over-digestion with exonuclease III.

④ The conditions of Mung bean nuclease digestion have been carefully determined. The use of higher temperatures leads to 'nibbling' by the enzyme with concomitant lowering of ligation efficiency and increased ratio of deletions with incorrect endpoints.

⑤ A titration of linker concentration is recommended.

- dCTP $\qquad$ 0.5 μl of 200 nM
- dTTP $\qquad$ 0.5 μl of 200 nM
- Sequenase $\qquad$ 3 U diluted to 1.5 U/μl with TE.①

3 Divide the reaction mixture from Step 2 into four 3.5 μl aliquots in fresh 1.5 ml microcentrifuge tubes for the subsequent base-specific reactions. To each tube add 2.5 μl of a base-specific α-thio-dNTP/dNTP mix. Incubate for 5 min at 37°C.②

4 Stop reactions by phenol extraction and ethanol precipitation in the presence of 20 μg tRNA carrier.③ Wash pellets in 70% ethanol.

5 Redissolve pellets in 100 μl of restriction enzyme digestion buffer and digest to completion with enzyme B (see *Figure 4*).

6 Stop reactions by phenol extraction, reprecipitate with ethanol,④ wash pellets with 70% ethanol.

7 Redissolve pellets in 90 μl of TE and add 10 μl of 10× exonuclease III reaction buffer. Add 80 U of exonuclease III and incubate at room temperature for 3 min.③

8 Stop reactions by phenol extraction, reprecipitate with ethanol,④ wash pellets three times with 70% ethanol.

9 Redissolve pellets in 45 μl of TE and add 5 μl of 10× Mung bean nuclease buffer then add 50 U of Mung bean nuclease. Incubate at 16°C for 5 min.④

Protocol 15. Generation of sequence-specific deletion mutants

10 Place the reactions on ice and stop by the addition of:
  - 1 M Tris-HCl (pH 9.0)   10 μl
  - 5 M LiCl   10 μl
  - 20% SDS   5 μl
  - distilled water   25 μl

  before phenol extraction and ethanol precipitation.[4] Wash pellets in 70% ethanol.

11 Ligate reaction products to 100 ng phosphorylated linkers⑤ at 16°C, in a 10 μl reaction. Following inactivation of the ligase, digest the DNA to completion with enzyme A and the enzyme that cuts the chosen linker (see *Figure 4*). Remove excess linkers by fractionation over a Schleicher and Schuell Elutip using conditions described by the manufacturers.

12 Ligate the reaction products to an empirically determined amount of vector backbone cut with enzyme A and the enzyme that cuts the chosen linker. Transform ligation mixes into *E. coli* (electroporation is recommended for this) and characterize recombinant clones.

Pause points

[1] Precipitates may be stored at 4°C for several days in this form.
[2] Phage may be stored at 4°C for several days in this form.
[3] Phage DNA may be stored indefinitely at −20°C.
[4] DNA may be stored indefinitely at −20°C.

Reagents

Absolute ethanol⚠

Acrylamide stocks (mono:bis at 20:1 ratio in 7 M urea)⚠

AMV reverse transcriptase

Brain–heart infusion broth

Chloroform⚠

Dimethyl sulfate (DMS)⚠

Distilled water

DMS buffer (50 mM sodium cacodylate, 1 mM EDTA pH 8.0)⚠

70% Ethanol⚠

Formic acid⚠

F positive *Escherichia coli* strain (e.g. JM101)

1 M HCl

Hydrazine

10× KS buffer (66 mM Tris-HCl pH 7.5, 66 mM MgCl₂, 500 mM NaCl, 20 mM DTT)

L broth (LB)

LB agar

2-Mercaptoethanol⚠

1.25 mM Mix of each dNTP

NaCl (solid)

2 M NaNO₂ (freshly made)

PEG-8000 (solid)

Phage-specific primer (50 pmol/µl)

Recombinant M13 phage containing fragment to be mutagenized

SM (5.8 g NaCl, 2.0 g MgSO₄.7H₂O, 0.1 g gelatin per liter in 50 mM Tris-HCl pH 7.5)

2.5 M Sodium acetate pH 4.3

3 M Sodium acetate pH 5.2

3 M Sodium acetate, pH 7.0

100 mM Sodium phosphate, 10 mM EDTA pH 7.0

10× TBE (108 g Tris-HCl base, 55 g boric acid, 7.5 g EDTA, make up to 1 liter with deionized water, autoclave, store at room temperature)

TE (10 mM Tris pH 7.4, 1 mM EDTA)

Tris-equilibrated phenol⚠

Yeast tRNA (10 µg/µl in TE)

Equipment

9 cm Bacterial petri dishes

Ice-water bath

37°C Incubator

Microcentrifuge

1.5 ml Microcentrifuge tubes

Preparative centrifuge (e.g. Beckman J6) and appropriate tubes
Set of micropipettes (e.g. Gilsons or Finn pipettes)
15 ml Snap-cap tubes
Shaking incubator and conical flasks (500 ml and 2 liter)

Spectrophotometer and cuvettes
Vortex apparatus
Water baths at 37°C, 40°C, 90°C and 95°C
Yellow and blue polypropylene tips for micropipettes

Procedure

Large-scale preparation of ssDNA

1 Plate out the recombinant phage containing the insert to be mutagenized on a lawn of host *E. coli* (e.g. JM101), and leave at 37°C overnight. Pick a single, well isolated plaque, and grow in 3 ml of brain–heart infusion broth in a 15 ml snap-cap tube.

2 Whilst the recombinant phage culture is growing (see Step 1) inoculate 100 ml of brain–heart infusion broth in a 500 ml conical flask with a single colony of the host strain of *E. coli* (e.g. JM101). Grow with aeration at 37°C until the culture reaches an OD_{600} of 0.2. Inoculate this with 1 ml of the saturated phage culture and continue to grow at 37°C for 30 min.

3 Transfer the culture from Step 2 to 1 liter of brain–heart infusion broth in a 2 liter conical flask and grow overnight with aeration at 37°C.

4 Precipitate bacteria by centrifugation for 10 min at no less than 10 000 *g*. Discard pellet.

5 To supernatant, add PEG-8000 and NaCl to final concentrations of 4% and 0.5 M respectively. Allow to stand overnight at 4°C. [1]

Notes

Time required: 7 days

(1) See *Figure 5* for outline of method.
(2) All conditions give approximately 10% single-base substitutions in a 65 bp region. To increase the degree of mutagenesis, increase the incubation times appropriately.
(3) It is important to remove all traces of formic acid.
(4) It is important to remove all traces of hydrazine.

6 Precipitate phage by centrifugation for 20 min at no less than 10 000 *g*. Carefully remove supernatant. Resuspend pellet in 20 ml of SM. [2]

7 Extract with an equal volume of chloroform, and then extract with an equal volume of phenol.

8 Add 2 ml of 3 M sodium acetate (pH 5.2) then 50 ml of absolute ethanol. Store at $-20°C$ for at least 1 h. Recover single-stranded DNA by centrifugation at no less than 10 000 *g*. Wash pellet with 70% ethanol, and dissolve in 5 ml TE. Quantitate DNA recovery using an OD_{260} of 1.0 as equivalent to 36 μg/ml. [3]

Mutagenesis

1 Prepare a solution of 40 μg of single-stranded DNA in 100 μl reaction volume. This will suffice for five mutagenesis reactions, carried out as follows:

(i) **Formic acid treatment (A+G specific):** In a microcentrifuge tube, set up the following:
- DNA solution   20 μl
- distilled water   15 μl
- formic acid   65 μl

incubate at room temperature for 15 min, stop with 100 μl 3 M sodium acetate pH 7.0, 200 μl distilled water, 40 μg tRNA, 1 ml ethanol. [4] Mix, chill, centrifuge as normal. Reprecipitate samples once with an additional 20 μg tRNA, [4] and wash with 70% ethanol. (3) Redissolve final pellet in 25 μl TE, spin out insoluble debris, and freeze supernatant until required for primer extension.

Protocol 16. Single-stranded DNA mutagenesis

(ii) **Hydrazine treatment (C+T specific):** In a microcentrifuge tube, set up the following:

- DNA solution   20 μl
- distilled water   20 μl
- hydrazine   60 μl

incubate at room temperature for 15 min, stop with 100 μl 3 M sodium acetate pH 7.0, 200 μl distilled water, 40 μg tRNA, 1 ml ethanol. 4 Mix, chill, centrifuge as normal. Reprecipitate samples once with an additional 20 μg tRNA, ④ and wash with 70% ethanol. 4 Redissolve final pellet in 25 μl TE, spin out insoluble debris, and freeze supernatant until required for primer extension.

(iii) **DMS reaction (A specific):** In a microcentrifuge tube, set up the following:

- DNA solution   20 μl
- DMS buffer   80 μl
- DMS   1 μl

incubate at room temperature for 15 min, stop with 5 μl 2-mercapto-ethanol, 40 μl 3M sodium acetate pH 7.0, 300 μl distilled water, 20 μg tRNA, 1 ml ethanol. 4 Mix, chill, centrifuge as normal. Reprecipitate samples once with an additional 20 μg tRNA, 4 and wash with 70% ethanol. Redissolve final pellet in 50 μl TE, spin out insoluble debris. Add 6 μl 1 M HCl, leave on ice for 1 h, stop with 40 μl 3 M sodium acetate, 300 μl distilled water, 20 μg tRNA, 1 ml ethanol. 4 Mix, chill, centrifuge as normal. Reprecipitate samples once with an additional 20 μg tRNA, 4 and wash with 70% ethanol. Redissolve final pellet in 25 μl TE, spin out insoluble debris, and freeze supernatant until required for primer extension.

(iv) **DMS reaction (G specific):** In a microcentrifuge tube, set up the following:
- DNA solution   20 µl
- DMS buffer   80 µl
- DMS   1 µl

incubate at room temperature for 15 min, stop with 5 µl 2-mercaptoethanol, 40 µl 3 M sodium acetate pH 7.0, 300 µl distilled water, 20 µg tRNA, 1 ml ethanol. [4] Mix, chill, centrifuge as normal. Reprecipitate samples once with an additional 20 µg tRNA, [4] and wash with 70% ethanol. Redissolve final pellet in 50 µl TE, spin out insoluble debris. Add 6 µl 100 mM pH 7.0 Na_2PO_4/10 mM EDTA, incubate at 90°C for 15 min, stop with 40 µl 3 M sodium acetate, 300 µl distilled water, 20 µg tRNA, 1 ml ethanol. [4] Mix, chill, centrifuge as normal. Reprecipitate samples once with an additional 20 µg tRNA, [4] and wash with 70% ethanol. Redissolve final pellet in 25 µl TE, spin out insoluble debris, and freeze supernatant until required for primer extension.

(v) **Nitrous acid reaction (C+T specific):** In a microcentrifuge tube, set up the following:
- DNA solution   20 µl
- distilled water   20 µl
- 2.5 M sodium acetate, pH 4.3   10 µl
- 2 M $NaNO_2$ (fresh)   50 µl

incubate for 1 h at room temperature, stop with 40 µl 3 M sodium acetate

Protocol 16. Single-stranded DNA mutagenesis

pH 7.0, 300 μl distilled water, 20 μg tRNA, 1 ml ethanol.[4] Mix, chill, centrifuge as normal. Reprecipitate samples once with an additional 20 μg tRNA,[4] and wash with 70% ethanol. Redissolve final pellet in 25 μl TE, spin out insoluble debris, and freeze supernatant until required for primer extension.

Template extension reactions
1 To 25 μl of DNA sample treated in any of the above procedures, add:
- 10× KS buffer   10 μl
- phage specific primer   2 μl
- distilled water   63 μl

heat at 95°C for 5 min, then anneal at 40°C for 10 min. Add a further:
- 10× KS buffer   10 μl
- distilled water   70 μl
- dNTP mix (1.25 mM each dNTP)   20 μl
- AMV reverse transcriptase   50 U

incubate at 37°C for 1 h, recover DNA by phenol extraction and ethanol precipitation.[4] The mutated fragment can now be recovered from the reaction after restriction digestion with enzymes A and B and cloned into an appropriate vector backbone as outlined in *Figure 5*.

Pause points

[1] Precipitates may be stored at 4°C for several days in this form.
[2] Phage may be stored at 4°C for several days in this form.
[3] Phage DNA may be stored indefinitely at −20°C.
[4] Samples may be stored indefinitely at −20°C.

Reagents

Absolute ethanol⚠
Distilled water
2 mM dNTPs
10× PCR buffer [100 mM Tris-HCl pH 8.9, 500 mM KCl, 30 mM MgCl$_2$ 0.01% (w/v) gelatin]
Four PCR primers R1, R2, M and rev (20 mM)
Plasmid DNA containing the wild-type sequence

Taq DNA polymerase (5 U/μl)
Vector

Equipment

Agarose gel
Equipment and reagents for gel electrophoresis
Microcentrifuge
PCR thermal cycler

Procedure

Step 1

1 Perform two separate PCR reactions (the first with primer R1 and rev, the second with primers M and R2) in duplicate, containing:

- plasmid template   1 μl (20 ng)
- 10× PCR buffer (30 mM Mg^{2+})   5 μl
- dNTPs (2 mM)   5 μl
- primer R1 or M (20 mM) ①②   2 μl
- primer rev or R2 (20 mM)   2 μl
- *Taq* DNA polymerase (5 U/μl)   0.25 μl
- distilled water   34.75 μl.

Notes

Time required: 5 days

Day 1 - First PCRs <3 h; gel purification of PCR products <90 min.

Day 2 - Linear extension <90 min; final PCR <3 h; restriction digest overnight.

Day 3 – Gel purification of final product <90 min.

Day 3–5 – Cloning and sequencing.

① Primers R1 and R2 contain six 'stuffer' nucleotides followed by the restriction site (which may or may not be part of the wild-type sequence) and 15–20 nucleotides of wild-type promoter sequence. The stuffer preceding the

2 Use the programe: 94°C for 2.5 min (94°C for 1 min, 50°C for 1 min, 72°C for 2 min) 30×, 4°C. ③

3 Check for the correct size of fragments A (with primers R1 and rev) and B (using primers M and R2) on an agarose gel.

4 Purify all of the PCR products from agarose gels. ④ ☐1

Step 2

1 Estimate the DNA concentration of recovered fragments A and B.

2 Combine:
  * fragment A (~250 ng)   1.0 µl
  * fragment B (~250 ng)   1.0 µl
  * 10× PCR buffer (30 mM Mg^{2+})   2.5 µl
  * dNTPs (2 mM)   2.5 µl
  * *Taq* DNA polymerase (5 U/ml)   0.25 µl
  * distilled water   17.75 µl.

3 Anneal and extend using the programe: 94°C for 2.5 min (94°C for 1 min, 40°C for 1 min, 72°C for 2 min) 9×. ⑤

Step 3

1 To the same tubes add: ⑥
  * 10× PCR buffer (30 mM Mg^{2+})   2.5 µl
  * dNTPs (2 mM)   2.5 µl
  * primer R1 (20 mM)   2.0 µl

restriction site is necessary to allow efficient digestion by the restriction enzymes after the final PCR amplification.

② Primer M consists of 14–18 nucleotides of wild-type sequence (representing the 'anchor') followed by the mutated sequence (which may consist of single- or multiple-point mutations or complete mutation of 6–10 nucleotides) and another 15–30 nucleotides of wild-type sequence (to allow efficient hybridization). The length of this part of primer M should reflect the length of the mutated region. For single point mutations, 10 nucleotides or less of wild-type sequence should provide enough stability for PCR. Primer rev consists of the complementary anchor part of primer M (14–18 nucleotides).

③ The annealing temperature may have to be adjusted according to the T_m of the primers, bearing in mind the 5′ overhangs of primers R1 and R2 and the number of mismatches in primer M.

④ It is important to gel-purify fragments A and B to remove the wild-type plasmid and excess primers to ensure that all of the final PCR products contain the mutation(s).

⑤ Equal amounts (~ 250 ng) of each fragment are mixed with PCR buffer, dNTPs and *Taq* DNA polymerase, but without primers. A low annealing temperature is chosen to allow the two PCR fragments to anneal via their short, common overlapping anchor sequence, and the annealed fragments are extended with *Taq* DNA polymerase. Only heterodimers can be extended in this fashion, homodimers will be inaccessible and thus do not take part in this

- primer R2 (20 mM)   2.0 µl
- *Taq* DNA polymerase (5 U/ml)   0.25 µl
- distilled water   15.75 µl.

2 Amplify the extended heterodimers using the same PCR programe as in Step 1 (Section 2).

3 Ethanol precipitate the PCR fragment, redissolve in 50 µl distilled water ☐2 and perform the appropriate restriction enzyme digestion with enzymes R1 and R2.

4 After digestion, gel purify the DNA fragment and clone into a suitable vector.

5 The cloned fragment must be sequenced to ensure that only the desired nucleotides have been mutated. ⑦

reaction. This step is repeated nine times to increase the yield of extended heterodimers prior to PCR.

⑥ The extended heterodimer, comprising the whole mutated sequence is now amplified using the flanking primers R1 and R2. Using these two primers, only extended heterodimers can be amplified. Therefore, all the amplified fragment should contain the mutated sequence.

⑦ Each reaction (Steps 2 and 3) should be carried out in duplicate. This should yield enough mutated promoter fragment for purification and subsequent cloning.

Pause points

☐1 The purified DNA fragments can be stored indefinitely.
☐2 The redissolved DNA can be stored indefinitely.

VI PROTEIN–PROTEIN INTERACTIONS. P. A. Moore

Introduction

Integral to the formation and regulation of active transcription complexes is the ability of transcription factors to form protein–protein associations. These include the formation of multi-subunit complexes such as the RNA polymerase holoenzyme and the pre-initiation complex; the formation of homo- and heterodimeric specific DNA-binding transcription factor complexes; associations between transcriptional regulatory proteins and various components of the pre-initiation complex and holo-enzyme; and interactions between transcription factors and upstream regulatory proteins such as ligands and inhibitory molecules. A wide range of biochemical and genetic techniques are employed to study these protein associations which occur both on and off DNA. A major breakthrough in recent years in the study of protein–protein interactions has been the development of the yeast two-hybrid assay [1]. This genetic approach has been used in the molecular cloning of cDNAs encoding proteins that interact with known transcription factors [2,3] and for the subsequent mapping of sequences involved in protein interactions between proteins of known structure [4]. A biochemically based technique that has also proved useful in this respect is Far Western blotting, which involves the interaction between a radiolabeled protein and specific proteins within a mixture separated by SDS–PAGE and transferred to a filter. Both the yeast two-hybrid assay and Far Western blotting will be described in this chapter. In addition, the one-hybrid assay (an extension of the two-hybrid assay) which facilitates the genetic identification of proteins which interact with specific DNA sequences will be described.

The yeast two-hybrid system

An outline of the yeast two-hybrid assay approach towards the identification of associated proteins is presented in *Figure 7*. The assay takes advantage of the modular nature of transcription factors which have separable DNA-binding and transcriptional activation domains [1]. A cDNA encoding the protein, or stretch of protein, under investigation is translationally

(a) Transform yeast with bait plasmid (e.g. lexA:X)

(b) Co-transform yeast with target library (e.g. VP16:cDNA)

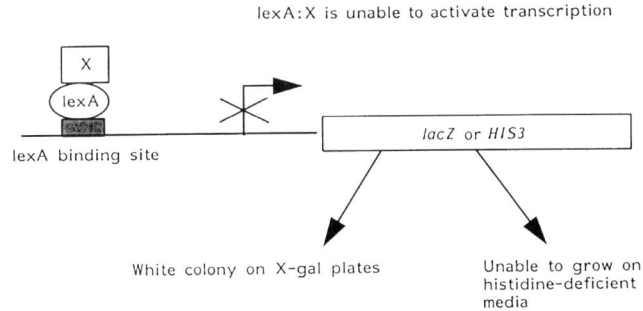

lexA:X is unable to activate transcription

lexA binding site

lacZ or *HIS3*

White colony on X-gal plates

Unable to grow on histidine-deficient media

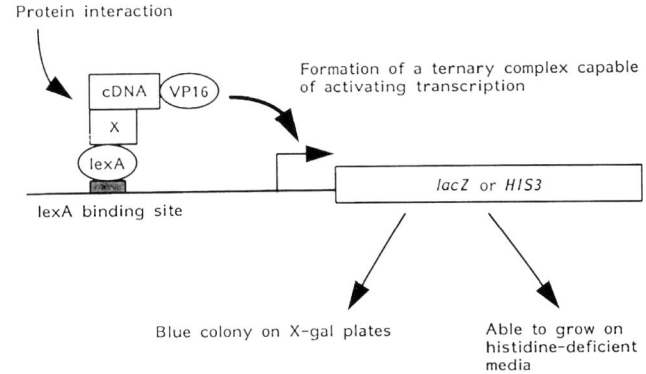

Protein interaction

Formation of a ternary complex capable of activating transcription

lexA binding site

lacZ or *HIS3*

Blue colony on X-gal plates

Able to grow on histidine-deficient media

Figure 7. The yeast two-hybrid system. (a) The protein of interest (X) is fused to the lexA DNA-binding domain and transformed into a yeast strain bearing a reporter gene(s) under the control of a promoter containing lexA DNA binding sites. (b) A cDNA expression library is introduced and an interaction between protein X and a cDNA-encoded target protein can result in activation of the selectable reporter gene.

fused in frame to a heterologous DNA-binding domain. This hybrid construct (referred to as the bait) is placed under the control of a yeast promoter in a selectable vector and transformed into yeast. The yeast strain transformed also contains a reporter gene whose expression is regulated by a promoter containing recognition sites for the DNA-binding domain present in the bait. Subsequently, a cDNA library fused to a transcriptional activation domain is transformed into the yeast strain expressing the bait protein. An association between any cDNA-encoded peptide (referred to as the target) and the bait will serve to bring a transcriptional activation domain into the proximity of the promoter. The formation of such a complex will activate transcription of the reporter gene, resulting in a selectable yeast phenotype. A range of controls are then performed to eliminate false positives, and the identified cDNA is isolated from yeast and subjected to further analysis.

Several considerations and preparations are required to be made, prior to utilizing the assay. Firstly, the bait expression vector needs to be constructed. While the DNA-binding domain of any protein can be used (including, if appropriate, the DNA-binding domain of the transcription factor under study), the DNA-binding domain of the yeast GAL4 protein or the *Escherichia coli* lexA protein are most commonly chosen. A major reason for this, is the availability of suitable yeast-reporter constructs bearing binding sites for either lexA or GAL4, upstream of reporter genes. A decision on whether a carboxyl-terminal or amino-terminal fusion of the DNA-binding domain to the transcription factor is constructed, is based on considering which will be least likely to disrupt the function of the transcription factor being studied. The choice of the reporter-gene promoter utilized is based on the DNA-binding domain present in the bait. The reporter genes most commonly used are the *lacZ* gene from *E. coli* [3], or a yeast amino acid biosynthetic gene such as *HIS3* [5,6]. Activation of the *lacZ* gene can be selected through a color-based assay, while *HIS3* activation can be measured by the ability to render an otherwise *his3* yeast strain capable of growing in the absence of histidine [7]. A yeast strain bearing both reporter systems is the most powerful, as it can be used to screen genuine protein associations more quickly. It is also preferable to utilize a yeast strain which bears the reporter genes stably integrated, and it is essential that the strain contains sufficient genetic markers to allow multiple plasmid selections. Finally, a choice has to be made regarding the cDNA library utilized. The library used will depend on where the bait is normally expressed, and its functions. While any transcriptional activation domain capable of activating transcription in yeast can be utilized, most commonly the cDNA is fused to the VP16 activation domain [3] or to the yeast GAL4 activation domain [1]. If a target cDNA library regulated by a galactose-inducible promoter is utilized, then a yeast strain bearing a functional *GAL4* gene is required, complicating the use of a GAL4-based reporter system.

The yeast one-hybrid system

The yeast one-hybrid assay extends the scope of the two-hybrid assay to allow isolation of sequence-specific DNA-binding proteins. Much of the experimental procedures, solutions and reagents utilized in this assay overlap with those used in the two-hybrid assay. The key differences in the one-hybrid assay are that the bait is a DNA sequence and only the cDNA–activation domain hybrid protein is expressed. Several novel DNA-binding proteins have been isolated using this technique including the neuronal transcription factor Olf-1 [14], and the B-cell factor Bob1 [15].

Protocols provided

References

1. Fields, S. and Song, O. (1989) *Nature*, **340**:245–246.
2. Hardy, C.F., Susse, L. and Shore, D. (1992) *Genes Dev.* **6**:801.
3. Dalton, S. and Triesman, R. (1992) *Cell*, **68**:597.
4. Ohana, B., Moore, P.A., Ruben, S.M., Southgate, S.M., Green, M.R. and Rosen, C.A. (1993) *Proc. Natl Acad. Sci. USA*, **90**:138.
5. Harper, J.W., Adami, G.R., Wei, N., Keyomarsi, K. and Elledge, S.J. (1993) *Cell*, **75**:805.
6. Zervos, A.S., Gyuris, J. and Brent, R. (1993) *Cell*, **72**:223.
7. Moore, P.A., Ruben, S.M. and Rosen, C.A. (1993) *Mol. Cell. Biol.* **13**:1666.
8. Ma, J. and Patshne, M. (1987) *Cell*, **51**: 113.
9. Chien, C.-T., Bartel, P.L., Sternglanz, R. and Fields, S. (1991) *Proc. Natl Acad. Sci. USA*, **88**:9578.

10. Durfee, T., Becherer, K., Chen, P.-L., Yeh, S.-H., Kilburn, A.E., Lee, W.H. and Elledge, S. (1993) *Genes Dev.* **7**:555.

11. Guthrie, C. and Fink, G. (eds) (1991) in *Methods in Enzymology,* Vol. 194. Academic Press, San Diego.

12. Sambrook, J., Fritsch, E.F. and Maniatas, T. (1989) *Molecular Cloning: a Laboratory Manual.* Cold Spring Harbor Laboratory Press, Cold Spring Harbor, NY.

13. Kaelin, W.G., Jr, Krek, W., Sellers, W.R., DeCaprio, J.A., Ajchenbaum, F., Fuchs, C.S., Chittenden, T., Blanar, M., Livingston, D. and Flemington, E.K. (1992) *Cell,* **70**:351.

14. Wang, M.M. and Reed, R.R. (1993) *Nature,* **346**:121–126.

15. Gsaiger, M., Knoepfel, L., Georgiev, O., Schaffner, W. and Hovens, C.M. (1995) *Nature,* **373**:360.

16. Jones, P. (1995) *Gel Electrophoresis: Nucleic Acids: Essential Techniques.* John Wiley & Sons, Chichester.

Reagents

Strains and DNA

Escherichia coli strain (eg. DH5α)

Saccharomyces cerevisiae strain containing an appropriate reporter gene and selectable auxotrophic markers for plasmid transformation (e.g. YJOZ [4], Y190 [5])

Target cDNA (1 mg) expression library DNA at a concentration of 2 µg/µl ①

Yeast bait (10 µg) expression plasmid DNA at a concentration of 0.1–1.0 µg/µl ②

Media and supplements

GAL.YNB (as GYNB but with 2% galactose replacing glucose)

GYNB (2% glucose, 0.67% yeast nitrogen base w/o amino acids, in distilled water)

GYNB plates (as GYNB with addition of 2% agar)

Selective GYNB contains the addition of amino acids (usually tryptophan, lysine, leucine and histidine) purines and pyrimidines (usually uracil and adenine) to satisfy the auxotrophic requirements of the yeast strain being used but omits the requirement provided by the plasmid marker gene.

Each supplement is added to 0.05 mg/ml. Stock solutions of each supplement are prepared at 10 mg/ml except uracil which is prepared at 5 mg/ml]

YPG 1% yeast extract 2% bactopeptone: 2% glucose, in water

X-Gal plates ③

Solutions

Carrier DNA [YEASTMAKER carrier DNA (Clontech)]

PEG [40% (w/v) polyethylene glycol 4000 (Sigma) prepared in TEL]

LacZ/X-Gal (27 µl 2-mercaptoethanol, 140 µl of X-Gal per 10 ml of Z buffer)

TE (10 mM Tris-HCl pH 7.6, 1 mM EDTA)

TEL (10 mM Tris-HCl pH 8.0, 1 mM EDTA, 100 mM lithium acetate)

TEN/SDS (100 mM NaCl, 10 mM Tris-HCl pH 8.0, 1 mM EDTA, 0.1% SDS)

X-Gal [25 mg/ml X-Gal (5-bromo-4-chloro-3-indolyl-β-d-galactopyranoside) prepared in dimethylformamide]

Z buffer (60 mM Na_2HPO_4, 40 mM NaH_2PO_4, 10 mM KCl, 1 mM $MgSO_4$)

Solvents

Chloroform △

Ethanol⚠
80% (v/v) Glycerol
Liquid N$_2$⚠
Phenol saturated with TE pH 8.0⚠
Sterile distilled water

Equipment

Acid washed glass beads (425–600 μm, Sigma)
Bench-top centrifuge capable of holding 50 ml centrifuge tubes
   and of reaching 1500 g
Floor centrifuge capable of holding 500 ml centrifuge bottles and
   reaching 3000 g
Microcentrifuge

24.5 cm × 24.5 cm Nitrogen membrane filters (e.g. Hybond-N,
   Amersham)
90 mm Petri dishes (20 cm × 20 cm)
Polystyrene ice-box suitable for holding liquid N$_2$
Shaking/rotary incubator at 30°C
Spectrophotometer and 1 ml disposable cuvettes
Sterile 50 ml, 500 ml centrifuge bottles
Sterile 250 ml, 2 liter culture flasks
Sterile 1.5 ml microcentrifuge tubes
Vortexing machine
30°C, 42°C, 50°C Water baths
Whatman 3MM paper

Procedure

A. Transformation of S. cerevisiae with bait plasmid

1 Inoculate 2 ml YPG with a single yeast colony, and incubate at 30°C with shaking until yeast reaches the stationary phase of growth (24–48 h).

2 Inoculate 50 ml of YPG with 50 μl stationary culture, and grow at 30°C overnight with shaking until the mid-logarithmic phase of growth is reached (OD$_{600}$ ~ 0.5; 1 × 10^7 cells/ml).④

3 Decant yeast culture into a 50 ml plastic centrifuge tube and harvest cells by centrifuging at 1500 g for 5 min in a bench-top centrifuge.

Notes

Time required: this assay takes approximately 3 weeks to complete

Section A – 3–4 days
Section B – 2–3 days
Sections C, D and E – 2–3 days for all three sections
Section F – 5–7 days
Section G – 3–4 days
Section H – 3–5 days

4 Wash cells twice with 10 ml sterile water and re-harvest cells as in Step 3.

5 Resuspend cells in 5 ml TEL to give a cell concentration of approximately 10^8 cells/ml.

6 Incubate yeast suspension at 30°C with gentle shaking for 1 h. [1]

7 Pipette 0.1–1.0 μg (in a volume of up to 10 μl) of the bait plasmid into a 1.5 ml microcentrifuge tube.

8 Add 300 μl of yeast cells from Step 6.

9 Add 700 μl of PEG, invert the tube several times, and place in a 30°C water bath for 1 h.

10 Transfer tube to a 42°C water bath for 15 min.

11 Microcentrifuge yeast–DNA mix at top speed for 5 sec. Wash cells twice in 0.5 ml sterile water, and resuspend in a final volume of 100 μl sterile water.

12 Pipette and spread 100 μl of yeast suspension on to a selective GYNB-Ura plate. Incubate at 30°C for 3–4 days until transformants appear. (5) [2]

B. Screening of bait transformants for reporter gene activation

1 Inoculate three bait transfomants separately into 2 ml GYNB-Ura and incubate at 30°C with shaking overnight.

2 Spot 5 μl of each overnight culture on to the appropriate selective GYNB

(1) For example cloned in pGAD1 [9] or pACT [10].

(2) For example cloned in pMA 424 [8] or pAS2 [5].

(3) Prepare the X-Gal plates as follows: mix 3.35 g of yeast nitrogen base (without amino acids) and 10 g agar in 300 ml distilled water, autoclave and allow to cool to 50°C in a water bath. Then add 50 ml of 0.7 M potassium phosphate pH 7.0 buffer (prepared by mixing 19.5 ml of 0.7 M KH_2PO_4 and 30.5 ml of 0.7 M K_2HPO_4), 50 ml of 20% glucose, the required amino acid and base supplements, and make the volume up to 500 ml with distilled water.

(4) The OD_{600} representative of a yeast strain in the mid-logarithmic phase of growth is strain dependent but usually lies between 0.3–0.6.

(5) For the sake of clarity it is assumed that the bait plasmid contains the *URA3* gene to allow selection in yeast, however, any selectable marker that is compatible with the host strain can be used.

(6) The reporter gene employed will dictate the selective agar used. For example if the reporter gene used is *HIS3*, then GYNB-His agar plates should be used. If the reporter gene is *lacZ* gene, then X-Gal plates should be utilized.

(7) If the *HIS3* gene is the reporter, addition of amino-triazole (AT; a competitive inhibitor of the *HIS3* gene product, available from Sigma) can block growth of the bait transformant on selective GYNB-His plates. The concentration of AT (1 mM–200 mM) required to block growth correlates with the ability of the bait protein to activate the *HIS3* reporter gene. Once conditions are

plate or X-Gal plate and as a control for growth, on to a GYNB-Ura plate. Incubate plates at 30°C for 2–3 days. ⑥

3 Score plates for growth and, if appropriate, formation of blue color on X-Gal plates.

4 Proceed to Stage C if the bait protein itself is incapable of activating transcription, demonstrated by the inability to render yeast blue on X-Gal plates or prototrophic for the biosynthetic reporter gene marker. ⑦

C. Transformation of target cDNA library

1 Inoculate 2 ml GYNB-Ura with a single bait yeast transformant and incubate at 30°C with shaking until the stationary phase of growth is reached (24–48 h).

2 Inoculate 500 ml YPG in a 2 liter culture flask with 500 µl of stationary culture and incubate at 30°C with shaking until the mid-logarithmic phase of growth (~0.5 OD_{600}) is reached.

3 Transfer cells to a 500 ml centrifuge bottle and harvest using a floor centrifuge at 3000 g for 5 min, resuspend pellet in 10 ml sterile distilled water and transfer to a 50 ml plastic centrifuge tube.

4 Harvest cells at 1500 g for 5 min using a bench-top centrifuge. Wash cell pellet once with 10 ml sterile distilled water and resuspend in 10 ml of TEL to give a cell concentration of 10^9 cells/ml.

5 Incubate yeast cell suspension at 30°C with gentle shaking for 1 h.

established that inhibit growth of the bait transformant on selective GYNB-His, then it is possible to proceed to Procedure C. It is necessary to continue adding the required dose of AT throughout the screening process but important not to use excessive amounts of AT that will block the identification of bait–target protein associations.

⑧ While any marker gene selectable in the host strain can be used to select for library co-transformants, for the sake of clarity it is assumed the *TRP1* selectable marker is employed.

⑨ Depending on the efficiency of transformation, Procedure C, Step 6 may have to be repeated several times to obtain full library representation.

⑩ The number of transformants obtained will depend on the efficiency of the transformation in Procedure C. For detection purposes, the maximum number of colonies on each plate should not be greater than 20 000.

⑪ The length of time it takes for blue colonies to appear will depend on the sensitivity of the assay and the strength of the interaction between the bait and target proteins.

⑫ If either the bait or the target library are regulated by a galactose-inducible promoter, the X-Gal plates should include 2% galactose.

⑬ The volume of cell suspension used should give an even spread of cells across the plate.

⑭ Cells can be washed off by pipetting a volume of 2 ml on to each plate, letting the plate shake gently on a platform shaker at room temperature for 30 min and then pipetting

6 Add 20 µg of the target cDNA plasmid library (containing the *TRP1* selectable genetic marker) and 100 µl of carrier DNA (10 mg/ml) to 10 ml yeast cells from Procedure C, Step 5, followed by 30 ml PEG. Invert tube several times and incubate in a 30°C water bath for 1 h. ⑧ ⑨

7 Transfer yeast–DNA mix to a 42°C water bath and incubate for 20 min.

8 Centrifuge at 1500 *g* for 5 min using a bench-top centrifuge at room temperature. Wash cells twice in 5 ml sterile distilled water, and resuspend in a final volume of 1 ml sterile distilled water.

If the initial selection screen is for activation of the *lacZ* reporter gene follow Stage D, and if it is for activation of an amino acid biosynthetic gene follow Stage E.

D. Screening for lacZ reporter gene activation

1 Dilute yeast cell suspension with sterile water to give a final volume of 100 ml.

2 Plate 5 ml of cells directly on to 20 separate Amersham Hybond-N filters (20 cm × 20 cm) laid on GYNB-Trp-Ura (24.5 cm × 24.5 cm) plates. Incubate plates at 30°C. Colonies will grow on the surface of the nylon membrane. ⑩

3 If expression of the target cDNA library and/or the bait are regulated by a galactose inducible promoter, the filters should be transferred to fresh plates containing GalYNB-Trp-Ura and incubated for a further 18 h to induce expression.

off yeast cell suspension.

⑮ Plate out a series of dilutions to determine the concentration of colony formation units in the glycerol stocks.

⑯ If the initial transformation gave rise to 100 000 transformants then plate out 500 000 colonies. It is assumed that the *HIS3* gene is the reporter gene being used.

⑰ The objective of this stage is to grow the co-transformant through sufficient generations to obtain mitotic segregants which have lost the bait plasmid.

⑱ Plate out 100 ml of yeast culture diluted 10^{-4}, 10^{-5} and 10^{-6} in sterile distilled water. When selecting for loss of a *URA3* based plasmid it is possible to counter select with 5-fluoro-orotic acid containing plates [11].

⑲ If utilizing the *lacZ* reporter gene, then analyze on X-Gal plates and if using a biosynthetic marker utilize the appropriate GYNB selective agar.

⑳ Transformants which demonstrate activity in the absence of the bait plasmid are not considered further, as the activation observed is mediated directly by the target cDNA and does not depend on the presence of the bait plasmid. This is probably caused by the protein encoded by the target cDNA binding to the promoter region of the reporter gene and activating gene transcription.

㉑ If no transformants are obtained using chemical transformation of *E. coli*, we have found electroporation gives a 10–100-fold higher number of transformants.

4 Using large forceps, lift the colony-containing filters from one edge and immerse them with colonies facing up into liquid nitrogen for 5 sec to permeabilize yeast cells. Allow filters to thaw and semi-dry for 2 min at room temperature (colony side up).

5 Layer filters on to Whatman 3MM paper, moistened with LacZ/X-Gal solution sufficient to wet the filter but not to flood the colonies. Incubate the filters for up to 2 h at 30°C to allow blue color to develop.⑪

6 Positive (blue) colonies are picked (a small percentage of cells are viable) and streaked on to selective GYNB. Plates are incubated at 30°C for 2–3 days to obtain single colonies. The colonies are replica plated on to X-Gal plates and re-tested for their ability to turn blue. Positive single colonies are picked and streaked on to fresh GYNB selective plates and subjected to further analysis – proceed to Stage F.⑫

E. Screening for amino acid biosynthetic gene activation

1 Dilute yeast cells (from Procedure C, Step 7) with sterile distilled water to give a final volume of 10 ml.

2 Plate out 0.5 ml of yeast cells on to 20 separate 24.5 × 24.5 cm selective GYNB-Trp-Ura plates.⑬

3 Incubate plates at 30°C for 2–3 days, and then wash off colonies in a total volume of 40 ml sterile distilled water.⑭

4 Vortex yeast cell suspension, add glycerol to a final concentration of 20% and store as 1.5 ml aliquots at −70°C.

5 Thaw one aliquot and check plating efficiency by plating out serial dilutions on to GYNB-Trp-Ura plates.⑮

6 Incubate plates for 2–3 days at 30°C and then determine the cell concentration of the glycerol stock by counting the number of individual colonies on a suitable dilution plate.

7 Plate out at a multiplicity of five, the original number of transformants identified on to 20 large GYNB-Trp-Ura-His plates. If the target cDNA library or bait are regulated by a galactose inducible promoter utilize selective GALYNB plates.⑯

8 Incubate plates at 30°C for 2–3 days, pick single colonies and patch on to fresh selective GYNB-Trp-Ura-His plates. Incubate at 30°C for 2–3 days, and utilize single colonies for future analysis.

F. Curing of yeast strain of bait plasmid

1 Inoculate 10 ml GYNB-Trp+Ura separately with each yeast co-transformant displaying gene activation.⑰

2 Incubate cultures at 30°C for 2–3 days (with shaking), plate out serial dilutions on to GYNB-Trp+Ura.⑱

3 Patch 50 individual colonies on to both GYNB-Trp-Ura and GYNB-Trp+Ura plates and re-incubate plates at 30°C for 2–3 days.

4 Yeast capable of growing on GYNB-Trp+Ura but not on GYNB-Trp-Ura plates have been cured of the bait plasmid. They are picked and re-tested

for their ability to activate transcription of the reporter gene on the appropriate media.(19)

5 Strains which have now lost the ability to activate the reporter gene are used for further study as activation is dependent on the presence of the bait plasmid.(20)

G. Rescue of activation domain fusion plasmid

1 Inoculate individual 'positive' yeast cured of the bait plasmid (from Procedure F, Step 5) into 1.5 ml of selective GYNB liquid and incubate at 30°C with shaking overnight.

2 Harvest cells by microcentrifugation at top speed for 5 sec.

3 Resuspend cells in 200 μl TEN/SDS.

4 Add sterile glass beads to just below the level of the liquid, and mix vigorously by vortexing for 1 min.

5 Add 200 μl phenol, vortex for 1 min, microcentrifuge at top speed for 5 min at room temperature and transfer the upper aqueous layer to a second microcentrifuge tube.

6 Add 200 μl 1:1 phenol/chloroform, vortex for 1 min, microcentrifuge at top speed for 5 min at room temperature and transfer the upper aqueous layer to a third microcentrifuge tube.

7 Precipitate DNA by adding 20 μl of 3 M sodium acetate and 400 μl of ethanol. Incubate at $-20°C$ for at least 1 h.

8 Microcentrifuge at top speed for 15 min at 4°C. Wash pellet in 80% ethanol and redissolve in 10 μl TE.

9 Transform *E. coli* by established techniques with 5 μl of rescued DNA [12].㉑

10 Prepare mini-plasmid DNA from *E. coli* transformants by established techniques [12] and check restriction digest pattern to confirm that the correct plasmid has been rescued [16].

H. *Elimination of false positives*

1 Prepare transformation competent yeast cells as in Procedure A, Steps 1-6.

2 Co-transform into the yeast strain 1 μg of the target cDNA plasmid isolated in Stage G separately with 1 μg of the bait plasmid, with 1 μg of a plasmid expressing only the DNA binding domain present in the bait plasmid, and with 1 μg of a plasmid expressing the DNA binding domain fused to a protein unrelated to the original bait.

3 Plate out transformants on to the appropriate selective agar plates and incubate at 30°C for 3–5 days.

4 Pick at least three individual colonies from each co-transformation and patch them on to the appropriate media to determine if the reporter gene is being activated.

5 Proteins encoded by target cDNAs which are capable of activating the reporter gene only in the presence of the bait are studied further.

Pause points

1 Yeast at this point can be left overnight at 4°C.
2 Yeast transformants can be stored on agar plates sealed with plastic film at 4°C for up to 2 weeks and retain viability. For longer storage, 1 ml of a yeast transformant liquid culture should be stored in 20% glycerol at −70°C.

Protocol 18. The yeast two-hybrid assay

Reagents

As for *Protocol 18*

Equipment

As for *Protocol 18*

Procedure

1 Clone the DNA sequence under study into the context of a yeast promoter deleted of its UAS sequences (e.g. GAL1, CYC1) located upstream of a reporter gene such as *lacZ* or *HIS3*.

2 Transform this construct integratively into yeast [11].

3 Assay for activation of the reporter gene as described for the two hybrid assay (*Protocol 18*, Procedure B). If no endogenous activation is observed, a cDNA-activation domain fusion library is transformed into the yeast strain and activation of the reporter gene screened for as described for the two-hybrid assay.

4 Rescue plasmids from positive transformants (*Protocol 18*, Procedure G) and re-transform them into yeast strains bearing an integrated reporter gene under the control of the basal promoter either containing or lacking the DNA sequence under study. Only cDNA–activation domain fusion proteins which activate the reporter gene in the presence of the DNA sequence are considered for further study.

Notes

Time required
Step 1 – 1–2 weeks
Step 2 – 2 weeks
Step 3 – 2–3 weeks
Step 4 – 1–2 weeks.

Reagents

0.25 mg/ml Aprotinin

$[\gamma^{32}P]$-ATP (5000 Ci/mmol, 185 TBq/mmol, 10 mCi/ml, 370 MBq/ml)

50 mM Benzamidine

Bovine heart protein kinase A, resuspended in 40 mM DTT(Sigma)

100 mg/ml BSA

Guanidine hydrochloride

HBB (25 mM Hepes, 25 mM NaCl, 50 mM $MgCl_2$, 10 mM DTT pH 7.7)

Hyb-75 (20 mM Hepes, 75 mM KCl, 0.1 mM EDTA, 2.5 mM $MgCl_2$, 1 mM DTT, 1% nonfat dried milk, 0.05% Nonidet P40 pH 7.7)

10× Labeling buffer (200 mM Tris-HCl, 1 M NaCl, 120 mM $MgCl_2$, pH 7.6)

1.5 mg/ml Leupeptin

MTPBS buffer (150 mM NaCl, 16 mM Na_2HPO_4, 4 mM NaH_2PO_4, pH 7.3)

Nonfat dried milk (e.g. Marvel)

Nonidet P40

0.25 mg/ml Pepstatin

25 mM Phenylmethylsulfonylfluoride (PMSF)⚠

Protein molecular weight markers

Radiolabeled protein①

SDS–PAGE mixes and buffers

Sterile deionized water

Equipment

Centricon-10 filters (Amicon Corporation)

Electroblotting apparatus (e.g. BioRad mini trans-blot electrophoretic transfer cell)

Immobilon membranes (Amicon Corporation)

Microcentrifuge

Micropipettes (e.g. Gilsons or Finn pipettes)

Perspex radiation shield

Plastic film(Dow)

Polythene bags and bag sealing apparatus

Preparative low-speed centrifuge (e.g. Beckman J6)

1.5 ml RNase-free microcentrifuge tubes

RNase-free yellow and blue polypropylene tips for micropipettes

SDS–PAGE apparatus

Shaking platform

Vortex apparatus

Water bath set at 37°C

Western blot

Procedure

Radiolabeling of recombinant protein

1 At room temperature, add to a 1.5 ml microcentrifuge tube in the following order:

- sterile deionized water to a final volume of 50 μl x μl
- 10 labeling buffer   5 μl
- $[\gamma^{32}P]$-ATP (100 μCi, 3.7 MBq)   10 μl
- 25 mM PMSF   1 μl
- 50 mM benzamidine   1 μl
- 1.5 mg/ml leupeptin   1 μl
- 0.25 mg/ml aprotinin   1 μl
- 0.25 mg/ml pepstatin   1 μl
- recombinant protein (1)   1–20 μg
- catalytic subunit from bovine heart protein kinase A   20 U

incubate at 37°C for 1 h.

2 Add 1 μl of BSA stock, and then add 2 ml of MTPBS buffer containing 40 μl each of PMSF, benzamidine, leupeptin, aprotinin and pepstatin stocks.

3 Remove unincorporated $[\gamma^{32}P]$-ATP by centrifuging through a Centricon-10 filter at 2500 g and 4°C. Add a further 2 ml of MTPBS buffer containing protease inhibitors, and repeat the centrifugation. (2) 1

Far Western blotting

1 Resolve proteins from cell lysates by SDS–polyacrylamide gel electrophoresis (SDS–PAGE) using standard 'stacked' gels as described

Notes

Time required: 3 days

(1) Until about 5 years ago, the only effective way to label proteins *in vitro* utilized iodination. This approach is unsatisfactory due to the high levels of isotope required, and relatively low specific activities obtained, and has been replaced by two alternatives: (i) ^{35}S or ^{14}C are incorporated into the protein of interest by direct synthesis using an *in vitro* translation system; or (ii) ^{32}P is incorporated into the protein of interest using specific protein kinases. The commonest application uses protein kinase A, and the consensus phosphorylation site is usually introduced into the protein by constructing a fusion protein in an appropriate vector (e.g. pGEX-2TK [13]). Alternatively, endogenous kinase consensus sites can be used where they are available. The protocol here describes *in vitro* phosphorylation using protein kinase A.

(2) This procedure generates protein in a volume of less than 300 μl labeled to a specific activity of at least 10^7 c.p.m./ml.

(3) See *Protocol 3*.

(4) This step can be omitted if it is not required to denature the target protein before incubation with the recombinant probe.

in Sambrook *et al.* [12]. Protein mobilities are assessed by comparison to protein molecular weight markers.

2 Transfer gel on to Immobilon membranes by electroblotting using the manufacturer's instructions for the electroblot apparatus. ③

3 Block filter overnight at 4°C in 200 ml HBB + 5% nonfat milk + 0.05% Nonidet P40 with gentle shaking.

4 Immerse filter in 500 ml of HBB + 6 M guanidine hydrochloride and rock for 15 min with one change of solution. Transfer the filter to 500 ml of HBB + 3 M GHCl and rock for 10 min. Repeat the process with successive twofold dilutions of GHCl in HBB until the concentration of GHCl is less than 0.2 M. Wash filter briefly in HBB and then re-block in HBB with 5% nonfat milk and 0.05% Nonidet P40. ④

5 Wash filter twice in 500 ml Hyb-75 for 10 min.

6 Transfer filter to a sealable polythene bag. Add 50 ng/ml of radiolabeled recombinant protein in Hyb-75 solution (as little as 5 ml will be adequate). Seal bag and incubate filter overnight at 4°C.

7 Wash filter three times in 500 ml of Hyb-75 for 5 min each with gentle shaking.

8 Air dry filter, wrap in plastic film and autoradiograph.

Pause point

1 Phosphorylated protein can be stored at 4°C overnight.

Protocol 20. Far Western blotting

VII   PHOSPHORYLATION OF TRANSCRIPTION FACTORS.   A. R. Clark

Introduction

Protein phosphorylation is a rapid and efficient means of altering the properties of several transcription factors in response to extracellular signals [1–4]. Such modification may affect the subcellular localization of a transcription factor, its DNA binding properties or its ability to activate transcription. A paradigm for phosphorylation-regulated transcriptional activation is the cyclic AMP response element-binding factor CREB, which is phosphorylated at serine 133 by the cAMP-dependent protein kinase A [5], or by calmodulin- or nerve growth factor-induced kinases [6,7]. This increases the affinity of CREB for CREB-binding protein (CBP), which has the properties of a transcriptional cofactor, and may contribute to the transcriptional activation signal transmitted to the polymerase complex [8–11]. Since the pioneering work on CREB, the number of transcription factors known to be regulated by phosphorylation has grown rapidly. Several of these factors lie at the end of multi-step phosphorylation cascades, which in some cases are now reasonably well understood.

Expression of recombinant transcription factors in *Escherichia coli*

The elucidation of signal transduction pathways is beyond the scope of this chapter. It is assumed that reporter gene assays and analyses of protein–DNA interactions have implicated a particular factor in a transcriptional response to extracellular signal(s). The next question is whether this factor is regulated at the level of phosphorylation. For *in vitro* phosphorylation studies it is necessary to purify substantial quantities of the transcription factor in an unmodified state. The most straightforward way is to synthesize the protein in *E. coli* as a fusion product with the bacterial glutathione S transferase (GST) protein, followed by purification using glutathione-Sepharose beads (*Protocol 21*). This system is rapid and efficient, usually generating large quantities of protein in a nondenatured state. Cloning vectors are readily available; those supplied by Pharmacia contain multiple cloning sites in all three translational frames, and encode cleavage sites for thrombin or factor Xa proteinases downstream of the GST coding sequence, so that the transcription factor moiety can be cleaved from the GST moiety following purification. Alternatively, substrate protein can be easily eluted from beads as a GST fusion protein. In practice, it is often easiest to perform *in vitro* phosphorylation assays 'in batch', that is, with the GST fusion protein still immobilized on beads. Other methods of protein purification such as polyhistidine tagging share many of the advantages listed above, but are not described here.

Mapping of phosphorylation sites in transcription factors

The first step is to clone all or part of the transcription factor coding sequence into a GST expression vector. Very large GST fusion proteins tend to be synthesized less efficiently, thus it may be worth cloning subregions of the transcription factor, in addition this approach makes it easier to locate potential phosphorylation sites. Recombinant clones are identified by restriction analysis, and fusion protein synthesis is checked by small-scale purification, then a medium-scale preparation is performed to obtain sufficient material for several *in vitro* kinase assays (*Protocol 22*). Typically, initial assays will be performed with whole cell extracts of cells treated under conditions known to stimulate transcriptional activation. It is important to investigate several time points, since stimulation of kinase activity may be very transient, and may occur in anything from a minute to several hours. Having established that the factor is subject to phosphorylation, deletion or point mutagenesis of the GST fusion protein expression vector can be used to determine which residues of the substrate are required for phosphorylation. Actual phospho-acceptor sites can be defined more precisely by phospho-amino acid analysis and phosphopeptide mapping (*Protocols 24* and *25*).

Characterizing kinases involved in the phosphorylation of transcription factors

Several approaches can be taken in attempting to identify the kinases responsible for transcription factor phosphorylation. The in-gel kinase assay (*Protocol 23*) can be used to identify the molecular weights of kinases active against the substrate of interest, however not all kinases remain active throughout the denaturation/renaturation procedure. Constitutively active and dominant-negative mutants have been identified for the components of some signal transduction pathways, and can be used in transfection experiments, in combination with transcriptional assays, or with kinase assays. Purified preparations of numerous kinases are commercially available. These may be used for *in vitro* kinase assays, bearing in mind that high concentrations of kinase may phosphorylate nonphysiological substrates, so that it is necessary to titrate kinase activity carefully, and to perform comparisons with a known, authentic substrate. The baculovirus system may also be used to overexpress kinases in insect cells, from which they can subsequently be extracted. Animal tissue may be used as a convenient source of large amounts of material for biochemical characterization and/or purification of kinases. Finally, it may be possible to immunoprecipitate active kinases from appropriately treated cells. Antibodies to numerous cellular kinases are commercially available, but do not invariably work well in immunoprecipitated kinase assays.

Protocols provided

21. *Generation of GST fusion proteins*
22. In vitro *phosphorylation assay*
23. *In-gel kinase assay*
24. *Phospho-amino acid analysis*
25. *Phosphopeptide mapping*

References

1. Jackson, S.P. (1992) *Trends Cell Biol.* **2**:104.
2. Hunter, T. and Karin, M. (1992) *Cell*, **70**:375.
3. Karin, M. (1994) *Current Opin. Cell. Biol.* **6**:415.
4. Hill, C.S. and Treisman, R. (1995) *Cell*, **80**:199.
5. Gonzalez, G.A. and Montminy, M.R. (1989) **59**:675.
6. Sheng, M., Thompson, M.A. and Greenberg, M.E. (1991) *Science*, **252**:427.
7. Ginty, D.D., Bonni, A. and Greenberg, M.E. (1994) *Cell*, **77**:713.
8. Chrivia, J.C., Kwok, R.P.S., Lamb, N., Hagiwara, M., Montminy, M.R. and Goodman, R.H. (1993) *Nature (Lond.)*, **365**:855.

9. Kwok, R.P.S., Lundblad, J.R., Chrivia, J.C., Richards, J.P., Bachinger, H.P., Brennan, R.G., Roberts, S.G.E., Green, M.R. and Goodman, R.H. (1994) *Nature (Lond.)*, **370**:223–226.

10. Arany, Z., Newsome, D., Oldread, E., Livingston, D.M. and Eckner, R. (1995) *Nature (Lond.)*, **374**:81–84.

11. Parker, D., Ferreri, K., Nakajima, T., LaMorte, V.J., Evans, R., Koerber, S.C., Hoeger, C. and Montminy, M.R. (1996) *Mol. Cell. Biol.* **16**:694–703.

12. Gersten, D.M. (1996) *Gel Electrophoresis: Proteins: Essential Techniques.* John Wiley & Sons, Chichester.

13. Hibi, M., Lin, A., Smeal, T., Minden, A. and Karin, M. (1993) *Genes Devel.* **7**:2135–2148.

14. Cooper, J.A., Sefton, B.M. and Hunter, T. (1983) *Methods Enzymol.* **99**:387–402.

15. van de Geer, P., Luo, K., Sefton, B.M. and Hunter, T. (1993) in *Protein Phosphorylation: a Practical Approach* (D.G. Hardie, ed.), pp. 31–85. IRL Press, Oxford.

16. Sambrook, J., Fritsch, E.F. and Maniatas, T. (1989) *Molecular Cloning: a Laboratory Manual.* Cold Spring Harbor Laboratory Press, Cold Spring Harbor, NY.

Phosphorylation of transcription factors

Reagents

0.1% (w/v) Bromophenol blue
100 mM DTT (add immediately before use)
Factor Xa cleavage buffer (50 mM Tris-HCl pH 8.0, 1 mM CaCl$_2$)
GST elution buffer (50 mM Tris-HCl pH 8.0, 5 mM reduced glutathione). Make up fresh immediately before use
10% (v/v) Glycerol
IPTG △ (100 mM in distilled water, filter sterilize)
L-Broth (LB)
NETN [0.5% (v/v) NP40, 20 mM Tris-HCl pH 8.0, 100 mM NaCl, 1 mM EDTA]
NETNM [NETN + 0.5% (w/v) dried milk]
NETNG [NETN + 75% (v/v) glycerol]
1× SDS-PAGE loading buffer

2% (w/v) SDS
Thrombin cleavage buffer (50 mM Tris-HCl pH 8.0, 2.5 mM CaCl$_2$)
50 mM Tris-HCl (pH 6.8)

Equipment

15 ml Centrifuge tubes
Microcentrifuge
1.5 ml Microcentrifuge tubes
Micropipettes
Pipette tips
Revolving wheel
SDS–PAGE apparatus
37°C Shaking incubator
Sonicator

Procedure

1 From a single fresh colony of *E. coli* expressing the GST fusion protein, inoculate an overnight culture in LB containing 100 mg/ml ampicillin. For control purposes, also inoculate a culture containing the parental GST expression vector, and process this in parallel with the recombinant.

Notes

Time required: 2 days

① This is a small-scale protocol suitable for initial screening of recombinant clones. For preparation of substrate, scale up to 40 ml or 400 ml of culture (Step 2).

② Try to avoid frothing. Start with a moderate power setting,

2 Inoculate 1/10 into 3 ml of LB containing 100 mg/ml ampicillin, and grow to an optical density at 600 nm of about 0.6 at 37°C with vigorous shaking.

3 Add IPTG to a final concentration of 1 mM, and continue to grow at 37°C for 3–4 h.

4 In the meantime, prepare the glutathione-Sepharose beads. These are supplied by Pharmacia as a 75% slurry in 20% ethanol, and are required as a 50% slurry in NETNM. Shake the bottle vigorously, transfer 1 ml of slurry to a 15 ml tube, add 10 ml of NETNM, shake, allow the beads to settle by gravity, then remove the supernatant. Repeat, and resuspend beads in 0.75 ml of NETNM. [1]

5 Transfer the *E. coli* to a 1.5 ml microcentrifuge tube and centrifuge for 5 min at 4°C in a microcentrifuge. All subsequent steps are performed at 4°C or on ice. Resuspend pellet in 300 μl NETN and transfer to a fresh microcentrifuge tube.

6 Disrupt cells by three brief (<10 sec) bursts of sonication, ② leaving on ice for at least 15 sec between bursts. Centrifuge the sonicate for 10 min in a microcentrifuge. Transfer supernatant to a fresh microcentrifuge tube.

7 Add 10 μl of glutathione-Sepharose beads from Step 4, vortex briefly, and mix on a revolving wheel for 1 h. Spin down beads in a microcentrifuge for 5–10 sec, then remove supernatant.

and gradually increase.

③ Liquid phase above the beads should not freeze.

④ Most protein should elute in the first step, however significant quanitities may also be found in the second and third eluates.

⑤ For instance the Bradford BioRad protein assay.

⑥ Thrombin or factor Xa may be removed by dialysis, however for most applications this is not necessary.

⑦ If yields of fusion protein are low, or if extensive proteolysis of the product has occurred, try one of the following: express proteins in a protease-deficient strain of *E. coli* (e.g. the TOPP strains supplied by Stratagene); induce fusion protein expression at 30°C rather than 37°C (Step 3); induce fusion protein expression overnight rather than for 3–4 h, using a final IPTG concentration of 0.1 mM (Step 3); express GST fusions containing smaller fragments of the protein of interest.

Protocol 21. Generation of GST fusion proteins

8 Add 1 ml NETN per tube, vortex briefly, and mix on a revolving wheel for 10 min. Spin down beads for 5–10 sec, remove supernatant, and repeat wash twice. For screening of small-scale preparations proceed to Step 12. For long-term storage, resuspend beads in an equal volume of NETNG, mix thoroughly, and store at $-20°C$.③ For elution of GST-fusion proteins from the beads, proceed to Step 9. For cleavage of the transcription factor moiety from the GST moiety, proceed to Steps 10 and 11.

9 Add elution buffer, mix gently, incubate on ice for 5 min, then centrifuge in a microcentrifuge for 5–10 sec. Remove eluate to a fresh microcentrifuge tube. Repeat elution three times, and analyze eluates as in Steps 10 and 12.④

10 Estimate yield of GST-fusion protein by Coomassie blue binding assay.⑤

11 Wash beads twice with 10 packed bead volumes of thrombin cleavage buffer or factor Xa cleavage buffer. For each ml of packed bead volume add 1 ml thrombin or factor Xa cleavage buffer containing the appropriate enzyme (10 mg/ml of fusion protein). Rotate or shake gently overnight at room temperature. Spin down beads in a microcentrifuge for 5 min, and transfer supernatant to a fresh microcentrifuge tube⑥ Proceed to Step 12.

12 Analyze by SDS–PAGE and Coomassie blue staining [12]. For screening of small-scale preparations, simply resuspend beads in 50 μl

of 1× SDS loading buffer, boil for 2 min and electrophorese a 10 μl aliquot. Otherwise, electrophorese approximately 1 μg of fusion protein, as judged by Bradford assay. To check for elution or cleavage of GST-fusion proteins, an aliquot of residual beads from Steps 9 or 11 should also be analyzed by SDS–PAGE. ⑦

Pause point

1 Once prepared, the beads may be kept at 4°C for up to a month.

Protocol 22. *In vitro* phosphorylation assay.① A. R. Clark

Reagents

0.1% (w/v) Bromophenol blue
100 mM DTT (add immediately before use)
Extraction buffer [400 mM KCl, 20 mM Hepes pH 7.4, 10 mM
  EGTA, 5 mM NaF△, 5 mM EDTA, 1 mM DTT△, 1 mM
  sodiun vanadate△, 10% (v/v) glycerol, 0.5% (v/v) Triton X 100,
  50 µg/ml PMSF, 5 µg/ml aprotinin, 5 µg/ml leupeptin, 5 µg/ml
  pepstatin, 0.1 µg/ml Okadaic acid△
10% (v/v) Glycerol
Kinase buffer (20 mM sodium β-glycerophosphate pH 7.4, 10 mM

MgCl₂ 5 mM NaF△, 2 mM DTT, 0.4 mM sodium vanadate△)
Phosphate-buffered saline (PBS)
2% (w/v) SDS
1× SDS-PAGE loading buffer
50 mM Tris-HCl (pH 6.8)

Equipment

Gel dryer
SDS–PAGE equipment
Water bath at 30°C

Procedure

1 An appropriate time after stimulation, wash cells twice with ice-cold PBS.
   Harvest by scraping in 1 ml ice-cold PBS, and spin down in a
   microcentrifuge for 1 min. Carefully remove all PBS.

2 By aspirating repeatedly with a yellow tip, resuspend cells in 200 µl of
   extraction buffer. Leave on ice for 5 min.② Spin down in a
   microcentrifuge at 4°C for 15 min to pellet cellular debris, and carefully
   remove supernatant to a fresh microcentrifuge tube. Use the Bradford
   assay to determine the protein concentration in the extract. Keep extract

Notes

Time required: 1 day

① This is a basic protocol for assaying kinase activity in
   crude whole cell extracts. The kinase assays are generally
   immobilized with substrate on glutathione-Sepharose
   beads, which works well even when the kinase is also
   immobilized (as in immune-complex kinase assays). If
   preferred, the substrate may be eluted or cleaved from
   the glutathione-Sepharose/GST complex, omitting Step 3
   in the protocol below. Adaptations to the extraction or

on ice, and proceed to kinase assay as soon as possible, since kinase activities may be unstable.

3 Wash substrate (approximately 2 μg per reaction, from *Protocol 21,* Step 8) twice in 1 ml of kinase buffer and resuspend in a small volume of kinase buffer.

4 In a total volume of 20 μl of kinase buffer combine 2 μg of substrate (from Step 3), 0.5 μl [γ^{32}P] ATP (5000 Ci/mmol, 185 TBq/mmol), cold ATP to a final concentration of 50 μM, ③ and cell extract from Step 2.④ Incubate at 30°C for 20 min, with gentle agitation every 5 min.⑤ Stop reaction by addition of 20 μl of 2× SDS–PAGE loading buffer. ☐1

5 Boil for 2–5 min, spin down in a microcentrifuge for 20 sec, then take 20 μl and subject to SDS–PAGE. Stain gel with Coomassie blue and destain to visualize substrate and confirm equal loading of tracks. Dry and expose to autoradiography film.⑥

assay buffers may be necessary in order to maximize the recovered kinase activity; however, we have found that this protocol works well for a variety of kinases and substrates.

② An additional sonication step is optional at this stage. Sonicate for only 5 sec, and use a low power setting.

③ The specific activity in this mixture is 5000 Ci/mol, but this can be adjusted upwards or downwards as appropriate. Specific activities ranging between 1000 and 50 000 are described in the literature.

④ As little as 1 μg of extract from UV-stimulated CHO cells is sufficient for strong phosphorylation of GST ATF2, however, it will be necessary to perform titrations to determine the optimum amount for any new substrate. For quantitative comparisons it is necessary to check the linear range of the assay. Under the conditions described, ATP is not limiting.

⑤ Unnecessary if substrate is soluble.

⑥ Smearing in the gel may be caused by inadvertent loading of glutathione-Sepharose beads, and is avoided by careful pipetting.

Pause point

☐1 The sample may be frozen at −20°C at this stage, but should be processed within a week.

Protocol 22. In vitro *phosphorylation assay*

Reagents

$[\gamma\text{-}^{32}\text{P}]$ ATP 5000 Ci/mmol, 185 TBq/mmol
0.1% (w/v) Bromophenol blue
Buffer A (50 mM Hepes pH 7.4, 10 mM DTT)
Buffer B (50 mM Hepes pH 7.4, 2 mM DTT, 15 mM MgCl₂,
    0.1 mM EGTA, 0.1 mM sodium vanadate⚠)
Buffer C [5% (w/v) TCA, 1% (w/v) NA₃PO₄]
Buffer A containing 20% (v/v) isopropanol
Buffer A containing 6 M guanidine hydrochloride
Buffer A containing 0.4% (w/v) Tween 20
Buffer B containing 10 mM ATP

100 mM DTT (add immediately before use)
10% (v/v) Glycerol
2% (w/v) SDS
1× SDS-PAGE loading buffer
50 mM Tris-HCl (pH 6.8)

Equipment

Gel dryer
Perspex radiation shield
SDS–PAGE apparatus
Shaking platform

Procedure

1 Prepare an SDS polyacrylamide gel containing 40 μg/ml of eluted GST fusion protein substrate (from *Protocol 21*, Step 9).

2 Prepare extracts from stimulated and unstimulated cells as in *Protocol 22*, Steps 1 and 2. Boil for 2 min in SDS–PAGE loading buffer, and subject to electrophoresis. ②

3 Trim excess gel material (stack, unloaded tracks). Shake the gel gently

Notes

Time required: 1 day

① In this assay, substrate is polymerized into the polyacrylamide gel, kinases are separated by electrophoresis, and allowed to phosphorylate the substrate in the gel following denaturation and renaturation steps. This allows the determination of the molecular weights of kinases which act upon the substrate. To demonstrate the specificity of for 1 h at room temperature in 250 ml of buffer A containing 20% v/v isopropanol. Remove liquid, and repeat.

4 Shake gently for 1 h at room temperature in 250 ml of buffer A. Remove liquid and repeat.

5 Shake gently for 1 h at room temperature in 100 ml of buffer A containing 6 M guanidine HCl. Remove liquid and repeat.

6 Soak gel at 4°C in 250 ml of ice-cold buffer A containing 0.04% v/v Tween 20. Change buffer at least four times over a period of up to 24 h. (3)

7 Shake the gel gently for 1 h at room temperature in 15 ml of buffer B.

8 Remove liquid and replace with 10 ml of buffer B containing 50 mM cold ATP and 100 µCi (3.7 MBq) [γ^{32}P] ATP. Incubate at room temperature for 1 h with gentle agitation, ensuring that the surface of the gel does not become dry.

9 Remove the radioactive buffer and dispose of according to local safety regulations. Wash the gel extensively with several changes of 250 ml of buffer C, allowing at least 30 min for each wash.(4) Monitor the radioactivity in each wash removed, and when background levels of radioactivity are achieved(5) dry the gel and expose to autoradiography film.(6)

phosphorylation reactions it is important to perform controls in which GST protein itself is polymerized into the gel, and to use extracts from both stimulated and unstimulated cells. It should be noted that some kinases will not remain active through the denaturation and renaturation process, and thus will not be detected in this assay. Further disadvantages of this protocol are that it employs large quantities of [γ^{32}P] ATP, and generates large volumes of contaminated wash buffer in the final stages.

(2) 10–100 µg of protein can be applied to the gel.
(3) May include an overnight incubation. Do not agitate during this renaturation step.
(4) May include an overnight incubation.
(5) May require around 2 liters of total wash.
(6) The most frequent problems with this protocol are excessive background, and failure to renature active kinases. If no kinase activity is detected, an alternative denaturation–renaturation procedure can be tried [13]. Problems with background are usually eliminated by careful washing (Step 9).

Protocol 23. In-gel kinase assay

Reagents

0.1% Bromophenol blue
ε-Dintrophenyl-lysine
100 mM DTT (add immediately before use)
10% (v/v) Glycerol
6 M HCl⚠
0.2% (w/v) Ninhydrin in acetone⚠
PAA electrophoresis buffer [8.3 ml formic acid (88%)⚠, 59.3 ml
  glacial acetic acid⚠, 3.3 ml pyridine⚠ (toxic and unstable,
  store under nitrogen), double-distilled water to 1 liter]
Phosphoserine, phosphothreonine and phosphotyrosine standards
2% (w/v) SDS
1× SDS-PAGE loading buffer
50 mM Tris-HCl (pH 6.8)
Xylene cyanol FF

Equipment

Electroblotting apparatus for transfer of gels to PVDF or
  Immobilon membranes (BioRad Trans-Blot semi-dry transfer
  cell or equivalent).
Heat block at 110°C.
Oven at 65°C
Polythene film
Rotary evaporator
SDS–PAGE apparatus
Thin layer electrophoresis apparatus, HTLE 7000 or similar (see
  ref. 14; available from CBS Scientific Inc.)
TLC plates (Merck or Kodak)
Whatman 3MM filter paper

Procedure

1 Phosphorylate 5 μg of substrate as in Step 4 of *Protocol 22*. Increase
  the specific activity in the phosphorylation reaction by decreasing
  the concentration of cold ATP ② and/or by adding more [γ^{32}P]
  ATP. ③

Notes

Time required: 1 day

① This protocol yields information on whether the targets of
  phosphorylation are serines, threonines, tyrosines, or a
  combination. In turn this gives guidance for further

2 Subject sample to SDS–PAGE, and transfer gel to a PVDF or Immobilon membrane, following the manufacturer's instructions for the membrane and transfer apparatus. Cover membrane with polythene film to keep it moist, mark with luminescent or radioactive paint, and expose to autoradiography film.

3 Using markers for alignment with film, excise the radioactive band from membrane. Transfer to a screw-cap microcentrifuge tube.

4 Add 100 μl of 6 M HCl, and boil at 110°C for 75 min.

5 Add 50 μl of double-distilled water, mix, transfer liquid to fresh microcentrifuge tube, and lyophilize in a rotary evaporator.

6 Redissolve in 100 μl of double-distilled water and lyophilize. Repeat.

7 Redissolve in 10 μl of PAA electrophoresis buffer containing 1 mg/ml phosphoserine, phosphothreonine and phosphotyrosine standards. (4) Make up marker solution containing 5 mg/ml ε-dinitrophenyl-lysine (yellow) and 1 mg/ml xylene cyanol FF (blue) in PAA electrophoresis buffer.

8 Mark an origin line 4 cm from one edge of a cellulose thin layer chromatography plate (Merck or Kodak). Spot 2 μl of each sample on to origin line (5) at least 3 cm apart and at least 4 cm from edges. Also spot a 2 μl aliquot of marker solution to monitor electrophoresis. Soak two sheets of Whatman 3MM paper with PAA electrophoresis buffer, remove excess buffer, and apply to the chromatography plate 5 mm from the

delineation of actual phospho-acceptor sites by site-directed mutagenesis, and may give some indication concerning the type of kinases involved.

(2) Not below 20 μM.

(3) Up to 20 μCi.

(4) Available commercially (e.g. Sigma).

(5) Carefully spot 0.5 μl at a time on to an origin point marked with a cross, using a hair-drier to dry the solution as it is applied.

(6) Samples migrate towards the anode.

(7) For relatively pure peptides phosphorylated *in vitro*, this method normally gives an adequate resolution of phospho-amino acids. For improved resolution the phospho-amino acids can be resolved by two-dimensional electrophoresis, first at pH 1.9, then at pH 3.5 [14,15].

Protocol 24. Phospho-amino acid analysis

origin line on either side, allowing the buffer to creep up to the origin line. Remove 3MM paper. There should be no excess buffer left on the plate.

9 Immediately transfer the plate to the electrophoresis chamber and subject to electrophoresis for 45 min at 20 mA. ⑥ Dry in an oven at 65°C.

10 Spray with 0.2% ninhydrin in acetone, and return to oven for 10–15 min to develop color and visualize phospho-amino acid standards. Expose to autoradiography film. Phosphoserine migrates furthest towards the anode, followed by phosphothreonine and phosphotyrosine. Free phosphate migrates in advance of phosphoserine, whilst some partially digested phosphopeptides may be detected between the origin and the phosphotyrosine spot. ⑦

Reagents

0.1% (w/v) Bromophenol blue
Chromatography buffer [375 ml *n*-butanol, 250 ml pyridine⚠
(toxic and unstable, store under nitrogen), 75 ml glacial acetic
acid⚠, 300 ml deionized water]
Digestion buffer (50 mM NH_4CO_3 pH 8.0)
100 mM DTT (add immediately before use)
pH 1.9 Electrophoresis buffer (50 ml formic acid⚠, 156 ml glacial
acetic acid⚠, 1794 ml deionized water)
10% (v/v) Glycerol
2% (w/v) SDS
1× SDS-PAGE loading buffer
50 mM Tris-HCl (pH 6.8)

Equipment

Chromatography chamber
Electroblotting apparatus for transfer of gels to PVDF or
Immobilon membranes (BioRad Trans-Blot semi-dry transfer
cell or equivalent)
Geiger counter
Immobilon or PVDF membranes
Oven at 65°C
SDS-PAGE apparatus
Thin layer electrophoresis apparatus. HTLE 7000 or similar (see
ref. 14; available from CBS Scientific Inc.).

Procedure

1 Phosphorylate GST fusion substrate *in vitro*, increasing the specific
activity in the reaction mix, as in *Protocol 24*, Step 1.

2 Subject sample to SDS-PAGE, transfer to PVDF or Immobilin membrane,
and excise band as in *Protocol 24*, Step 2.

Notes

Time required: 3 days

① In phosphopeptide mapping the GST fusion substrate is
phosphorylated *in vitro* and purified by SDS-PAGE then
proteinase digested to completion, electrophoresed in the
first dimension and chromatographed in the second

3 Transfer to a fresh, siliconized③ screw-cap microcentrifuge tube, and incubate in blocking solution at 37°C for 30 min. Wash five times with 1 ml of freshly made digestion buffer. Add 200 µl of digestion buffer.④

4 Add 2 µl of thermolysin⑤ (1 mg/ml in 0.1 mM HCl) and digest at 55°C overnight.

5 Add a further 1 µl of enzyme the following morning, and a further 1 µl in the evening. Continue incubation overnight, ensuring that the membrane slice is still fully submerged.

6 Transfer solution to a fresh, siliconized screw-cap microcentrifuge tube. Wash membrane with 200 µl of fresh deionized water, and pool with first eluate. Check that most counts have come off the membrane.

7 Lyophilize to complete dryness in a rotary evaporator. Resuspend in 100 µl of fresh deionized water, and lyophilize again. Repeat this step until no white residue is visible in the microcentrifuge tube. Resuspend in 10 µl of pH 1.9 electrophoresis buffer.

8 Prepare marker solution containing 5 mg/ml ε-dinitrophenyl-lysine in pH 1.9 electrophoresis buffer.

9 Using a soft pencil, mark two origin crosses on a cellulose TLC plate (Merck or Kodak) as shown in *Figure 8*.

10 Prepare a blotter by stitching or stapling together two 25 × 25 cm sheets of Whatman 3MM paper, and using a sharp cork-borer to cut two 1.5 cm diameter holes in the positions indicated in *Figure 8*.

dimension to generate a distinctive pattern of phosphopeptide spots (a 'fingerprint'). Subsequently actual phospho-acceptor sites within individual phosphopeptides can be determined by amino-terminal sequencing.

② A number of proteinases with differing specificities may be used, and the conditions for both electrophoresis and chromatography can be varied extensively. For discussion see refs 14 and 15.

③ See ref. 16.

④ It is important that the membrane slice remains submerged during the extensive digestion step. It may be necessary to cut the slice into smaller fragments.

⑤ For instance Calbiochem. This preparation contains Ca^{2+}, which is required for enzyme activity.

⑥ Monitor with a Geiger counter after each application.

⑦ The cellulose is very fragile at this stage, and the plate must be handled with great care.

⑧ In this time the marker dye will migrate approximately 5 cm. Actual electrophoresis times may need to be adjusted if phosphopeptides have over-run, or are not adequately resolved.

⑨ Complete drying is critical for the subsequent chromatography step. The TLC plate may be left overnight at room temperature.

11 Carefully spot sample on to its origin, 0.5 μl at a time, whilst drying with a hair-drier. For a good fingerprint you will need to apply at least 10–20 c.p.m.⑥ Apply 2 μl of marker solution to its origin.

12 Soak blotter in pH 1.9 electrophoresis buffer, and remove excess buffer. Carefully lay blotter on top of TLC plate so that the origin crosses lie at the centers of the two holes. Allow buffer to creep inwards to the origins, concentrating the samples. Remove blotter, remove any excess buffer from the TLC plate, and immediately transfer to the electrophoresis apparatus.⑦ Electrophorese for 10 min at 600 V, then for 40 min at 1000 V.⑧

13 Remove TLC plate from the electrophoresis chamber, and air dry completely.⑨

14 Line a chromatography tank with Whatman 3MM paper, add 250 ml of chromatography buffer, cover, and allow the chamber to equilibrate for about 30 min.

15 Place the TLC plate into the chamber and allow to develop until the solvent front is 2 cm from the top. Remove, air dry completely, and expose to autoradiography film.

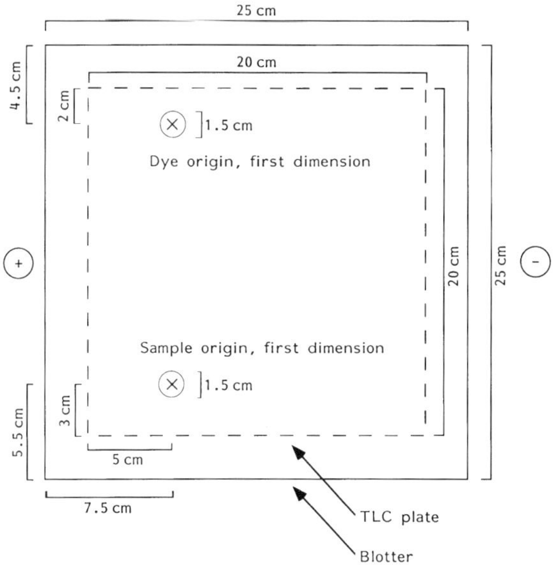

Figure 8. Layout of TLC plate and blotter for phosphopeptide mapping (adapted from van der Geer *et al.* 1993).

Protocol 25. Phosphopeptide mapping

VIII   HYPERSENSITIVE SITE MAPPING IN CHROMATIN.   J. Allan

Introduction

The DNA in eukaryotic cells is tightly packaged into chromatin and if genes contained within a chromatin fiber are to be transcribed, the organization of the chromatin must be modified. Surprisingly there is no substantial loss of core histones from a transcriptionally active gene and the elongation of RNA polymerase takes place on a nucleosomal template. Furthermore, the transition from a higher-order chromatin fiber to a more unfolded state appears to be transient, as when the elongation of RNA polymerase is inhibited, a solenoid-like, higher-order chromatin structure reforms over the previously, transcriptionally active region of the DNA [1].

The chromatin structure of a gene which is available for transcription or is being transcribed, can usually be distinguished from that of an inactive gene on the basis of two criteria, both of which relate to the sensitivity of the gene to the action of nucleases. Active genes display an increased, general sensitivity to DNase I and are about ten times more susceptible to the action of the nuclease when compared to the inactive form of the gene. This property is not specifically localized to the sequence of an active gene but is a feature of the chromatin domain containing the gene. It appears to reflect a general modification to the chromatin fiber, affecting many kilobases of DNA. Generalized DNase I sensitivity is usually taken to indicate unfolding of the higher-order chromatin fiber although the nature or extent of the unfolding is not understood [1].

Active genes frequently possess characteristically distinct chromatin sites which are hypersensitive to the action of nucleases [1–3]. Compared to bulk, surrounding chromatin, hypersensitive sites are generally two to three orders of magnitude more susceptible to nuclease cleavage. This exquisite sensitivity reflects specific and localized disruption to the higher-order chromatin fiber and is often caused by the binding of sequence-specific, nonhistone proteins to these regions. Hypersensitivity is associated with small regions of DNA no more than a few hundred base pairs in length. In some instances, hypersensitive sites are thought to reflect regions lacking a core histone octamer and therefore to be nucleosome-free, a situation likely to facilitate, or be a prerequisite for, the binding of nonhistone proteins to their DNA binding sites [3]. However, not all hypersensitive sites are nucleosome-free and in some cases the winding of DNA around the core histone octamer plays an important role in promoting protein–protein interactions within the complexes which form at hypersensitive sites [4].

Hypersensitive sites map to a variety of regulatory DNA elements which are responsible for controlling the expression of genes [1,3]. Thus promoters, enhancers, silencers and the elements which constitute locus control regions (LCRs) are hypersensitive when in an active context. Other elements on the chromatin fiber, such as the regions thought to constitute the boundaries of chromatin domains, may also display hypersensitivity [5]. It is worth remembering that LCR and domain boundary elements were identified as chromatin hypersensitive sites before a function was attributed to these DNA sequences. Indeed, it is now almost a rule of thumb that a nuclease hypersensitive site reflects a structurally distinct and, therefore, functionally important element within the chromatin fiber.

There is a general correlation between gene activity and hypersensitivity of associated regulatory regions. Hypersensitivity in the promoters of the linked β-globin genes, in humans or chickens for example, is strictly correlated to the temporal expression pattern of these genes. Similarly, the other regulatory elements (enhancers, LCRs) in these globin domains display hypersensitivity in a manner appropriate to the pattern of expression [1,3]. However, hypersensitive sites which are associated with permanent structural elements, such as chromatin domain boundaries, may be constitutively formed in all cell types [5]. Hypersensitive sites may also characterize the regulatory regions of genes which are prepared for activity but are not actually undergoing the process of transcription. For example, in an uninduced state, the heat shock genes in *Drosophila* are poised for transcription and contain a primed RNA polymerase molecule located in their promoter [4]. The particular architecture of these chromatin sites makes them hypersensitive to nucleases and DNase I hypersensitive sites were first identified in the promoters of these (inactive) genes [2].

General method

The general approach used to map hypersensitive sites has changed little from the method originally described [2,6]. Since then, substantial advances have been made in terms of characterizing the fine structure of hypersensitive sites, by the application of *in vivo* footprinting techniques, but here attention has been focused solely upon the basic method used to identify and locate hypersensitive sites.

In order that a region of the genome can be analyzed for hypersensitive sites, genomic clones covering the area, and partially characterized by restriction enzyme digestion, must be available. These will be needed to devise a strategy for mapping sites with respect to the (gene) sequence of interest and to prepare hybridization probes.

The various steps in the procedure are outlined in *Figure 9A*. Nuclei are first purified from cell lysates by centrifugation; note that the procedure described in *Protocol 26* is specific for chicken erythrocyte nuclei and that nuclei from other cell types will usually require a different isolation procedure. The purified nuclei are then treated with DNase I (*Figure 9A*(b)). Digestion should be to such an extent as to introduce an average of one cut within the region being mapped. The required conditions will probably vary as a function of the nuclei type and can only be determined experimentally. Typical conditions for DNase I digestion of avian erythrocyte nuclei are given in *Protocol 27*. After DNase I digestion, nuclear DNA is isolated and purified as described in *Protocol 28*. This DNA is then digested to completion with a restriction enzyme which has cleavage sites flanking the region being mapped (*Figure 9A*(c)), and again the DNA is purified (*Protocol 29*). The remainder of the method involves gel electrophoresis, blotting and filter hybridization, procedures which have been extensively described elsewhere [7] and are, therefore, not given in any detail here. Briefly, the DNA fragments are separated by electrophoresis in an agarose gel and transferred to a nylon or nitrocellulose filter [8]. Appropriate size markers should also be run on the gel. The filter is then hybridized [7] with a short DNA probe which is complementary to one end of the DNA sequences being mapped (*Figure 9A*(e)). This short probe can be isolated by restriction digestion of an appropriate genomic clone or may be amplified by PCR from that clone if the primers (and therefore sequence) are available. The hybridization probe is most effectively labeled by the use of random primers [9].

Figure 9B presents a schematic representation of the results one might expect from a hypersensitive site mapping. Autoradiography of the hybridized filter reveals the presence of a band with a size corresponding to the length of DNA between the

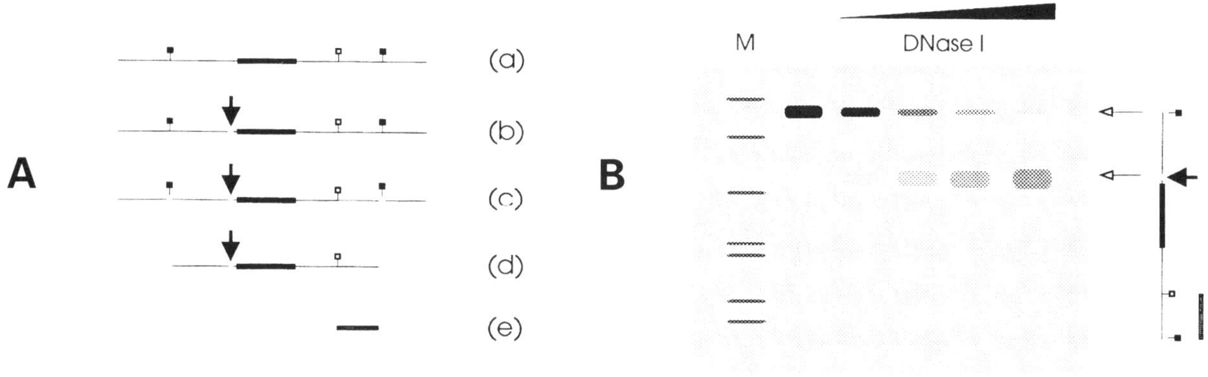

Figure 9. **(A)** Hypersensitive site mapping. A region of DNA containing a gene (▬▬▬) and flanked by restriction sites (■) (a) is cut with DNase I (↑) in nuclei (b) and, after purification, the DNA is cut to completion with the flanking restriction enzyme (c). Following electrophoresis of the DNA in an agarose gel and transfer to a filter, the digest is probed with a fragment of DNA complementary to one end of the region being mapped (e). **(B)** Schematic diagram showing the result of filter hybridization. The gel contains a DNA size marker (M) and a set of samples digested with increasing amounts of DNase I.

sites used for restriction enzyme digestion of the genomic DNA. This should be the only band present in the lane containing the control nuclear digest which lacked DNase I. During the course of digestion with DNase I this parent band should gradually decrease in intensity due to the accumulation of cleavages introduced by the nuclease. The presence of a hypersensitive site in the region being mapped will be indicated by the gradual appearance of a new band which is not present in the control digest. This band will often be rather diffuse as cleavage at the site may be spread over a 100–200 bp region. The size of the new, DNase I-dependent band, locates the hypersensitive site with respect to the restriction site which forms one end of the hybridization probe (*Figure 9B*).

Variations to the method

A variety of nucleases can be used to map hypersensitive sites. In addition to DNase I, DNase II and micrococcal nuclease, restriction enzymes can be employed [6,9]. Digestion conditions should be altered to accommodate the requirements of the different enzymes and, for restriction enzyme analysis, modification of the nuclei preparation may be beneficial [10].

For the initial identification of hypersensitive sites, the methods given here should suffice. However, if the fine structure of a previously identified site is to be studied in greater detail then high-resolution methods of analysis may be required. These approaches are essentially *in vivo* footprinting methods and usually employ primers and DNA amplification procedures (see Chapter 2).

Protocols provided

26. Preparation of chicken erythrocyte nuclei
27. Nuclease digestion
28. Purification of DNA
29. Restriction digestion

References

1. Wolffe, A.P. (1992) *Chromatin Structure and Function.* Academic Press, London.
2. Wu, C. (1980) *Nature* **286**:854.
3. Gross, D.S. and Garrard, W.T. (1988) *Ann. Rev.*

Biochem. **57**:159.

4. Thomas, G.H. and Elgin, S.C.R. (1988) *EMBO J.* **7**:2191.
5. Chung, J.H., Whiteley, M. and Felsenfeld, G. (1993) *Cell*, **74**:505.
6. Wu, C. (1989) *Methods Enzymol.* **70**:269.
7. McGhee, J.D., Wood, W.I., Dolan, M., Engel, J.D. and Felsenfeld, G. (1981) *Cell*, **27**:45.
8. Sambrook, J., Fritsch, E.F. and Maniatis, T. (1989) *Molecular Cloning: a Laboratory Manual.* Cold Spring Harbor Laboratory Press, Cold Spring Harbor, NY.
9. Feinberg, A.P. and Vogelstein, B. (1983) *Anal. Biochem.* **132**:6.
7. McGhee, J.D., Wood, W.I., Dolan, M., Engel, J.D. and Felsenfeld, G. (1981) *Cell*, **27**:45.
10. Caplan, A., Kimura, T., Gould, H. and Allan, J. (1987) *J. Mol. Biol.* **193**:57.

Hypersensitive site mapping in chromatin

Protocol 26. **Preparation of chicken erythrocyte nuclei.**① J. Allan

Reagents

Buffer A (0.25 M sucrose, 6 mM MgCl$_2$, 50 mM Tris-HCl pH 7.5, 0.5 mM EGTA and 0.2 mM PMSF)
Buffer A plus 1% (v/v) Triton X-100⚠
Chicken blood
Heparin sulfate solution
Phosphate-buffered saline (PBS) containing 0.5 mM ethylenebis tetra-acetic acid (EGTA), 0.2 mM phenylmethylsulfonyl fluoride (PMSF)⚠

Equipment

Aspirator
Centrifuge tubes
Glassware
Magnetic stirrer
Surgical gauze
Ultracentrifuge and rotors

Procedure

1 Collect chicken blood through two layers of gauze into ice-cold PBS containing 25 U/ml heparin sulfate.②

2 Pellet red blood cells by centrifugation at 2000 g for 5 min.

3 Remove supernatant and white cell layer (buffy coat) by careful aspiration.③

4 Wash red cell pellet with PBS a further two times.④

5 Resuspend red cell pellet in a minimum volume of buffer A.⑤

Notes

Time required: 1.5 hours

① The purification of nuclei from other sources will usually require modifications to the procedure described here.

② All steps in this protocol should be carried out at 4°C or on ice. The heparin sulfate prevents coagulation of the blood but need only be present in the first PBS buffer.

③ The white cell layer is like a skin on the red cell pellet and can easily be removed by aspiration with a pasteur pipette without too much loss of red blood cells.

6 Add dropwise to 10 vol. of buffer A containing 1% Triton X-100, stirring constantly. ⑥

7 Stir for 17 min.

8 Pellet nuclei by centrifugation at 2000 g for 5 min.

9 Wash nuclear pellet with buffer A a further two times. ⑦

④ Gently resuspend pellet in buffer by shaking and then centrifuge.

⑤ Gently resuspend pellet in buffer A by shaking and rotating the tube. It is important to ensure that the pellet is completely resuspended. The volume used here will determine the volume for cell lysis and subsequent centrifugation (Steps 6 and 7).

⑥ This step is most conveniently done in the cold room (4°C). Buffer A containing Triton should be in a beaker on a stirrer at a speed which ensures the rapid dispersion of the blood which is added slowly, drop by drop, from a pasteur pipette.

⑦ The pellet should be very gently resuspended by shaking and rotating the tube. Rough treatment may lead to nuclear lysis which results in foaming and stickiness at the surface of the nuclear suspension. A satisfactory preparation of clean erythrocyte nuclei should be a cream color.

Protocol 26. Preparation of chicken erythrocyte nuclei

Protocol 27. **Nuclease digestion.** J. Allan

Reagents

5 mM $CaCl_2$
Digestion buffer (0.25 M sucrose, 6 mM $MgCl_2$, 50 mM Tris-HCl pH 7.5, 0.1 mM EGTA and 0.2 mM PMSF)
50 U/ml DNase I in enzyme buffer
Enzyme buffer (50 mM Tris-HCl pH 7.6, 100 mM NaCl, 50 μg/ml BSA)
Erythrocyte nuclei
Termination buffer [20 mM EDTA, 0.4 % (w/v) SDS]

Equipment

Centrifuge tubes
Centrifuge and rotor
1.5 ml Microcentrifuge tubes
Microcentrifuge
Pipettes and tips
Water bath at 37°C

Procedure

1 Resuspend nuclei (see *Protocol 26*) in at least 10 vols of digestion buffer and remove duplicate 20 μl aliquots to determine DNA concentration. (1)

2 Pellet nuclei by centrifugation at 2000 *g* for 5 min.

3 Remove supernatant and resuspend nuclei at a concn 1 mg/ml (DNA) in digestion buffer.

Notes

Time required: 30 minutes

(1) The nuclei are washed into an appropriate buffer for nuclease digestion. Alternative buffers could be used for different nucleases (restriction enzymes). The DNA concentration is determined by diluting 20 μl of the suspended nuclei into 980 μl of 2 M NaCl, 5 M urea and shearing the mixture by sonication. After a brief spin in a microcentrifuge the optical density at 260 nm is read (1 OD_{260} = 50 μg/ml).

4 Prepare a set of microcentrifuge tubes containing 5 μl of 5 mM CaCl$_2$, a range of amounts of 50 U/ml DNase I (0, 0.5, 1, 2, 5 μl) and enzyme buffer to give a final volume of 10 μl. ②

5 Add 40 μl of nuclei to each tube, mix and incubate at 37°C for 5 min. ③

6 Add 50 μl of termination buffer to each tube and mix to stop reactions.

② A range of digestion conditions should be tested. Varying the nuclease concentration is the most effective way to determine optimum conditions. The extent of digestion can be assessed by inspecting the size distribution of the bulk DNA after electrophoresis in an agarose gel. Digestion may be carried out at lower temperatures to reduce the contribution of endogenous nucleases.

③ Before adding the nuclei to the nuclease–CaCl$_2$ mix, they should be equilibrated for a few minutes at the digestion temperature.

Protocol 28. **Purification of DNA.** J. Allan

Reagents

Absolute ethanol⚠
Buffer (0.5 M Tris-HCl pH 7.5) saturated phenol⚠ containing
 0.1 % hydroxyquinoline
Chloroform:isoamyl alcohol (24:1)⚠

70% Ethanol⚠
5 mg/ml Proteinase K
3 M Sodium acetate pH 5.2

Equipment

1.5 ml Microcentrifuge tubes
Microcentrifuge
Pipettes and tips
Vortex mixer
Water bath at 37°C

Procedure

1 Add proteinase K to the lysed nuclei (see *Protocol 27*) to a final
  concentration of 100 µg/ml. Incubate at 37°C, overnight.

2 Add an equal volume of phenol and vortex to form an emulsion.①

3 Centrifuge for 3 min to separate the organic and aqueous phases.

4 Remove the upper, aqueous phase with a pipette and transfer to a clean
  tube.②

5 Repeat Steps 2–4 twice.

6 Add an equal volume of chloroform and vortex to form an emulsion.

Notes

Time required: 1–2 hours

① **Caution:** Phenol is corrosive and will cause severe
  burning to skin and clothing. Wear gloves, protective
  clothing and safety glasses while working with phenol.

② The interphase between the aqueous and organic layers
  should be compact. Do not disturb this skin-like layer
  when removing the aqueous phase. The DNA in the
  aqueous phase will be viscous and pipetting can be
  improved by cutting off the ends of the plastic tips to
  create a bigger orifice.

③ Alternatively place in dry ice for 5 min or at −70°C for 30
  min.

7 Centrifuge for 3 min in a microcentrifuge to separate the organic and aqueous phases.

8 Remove the upper, aqueous, phase with a pipette and transfer to a clean tube.

9 Repeat Steps 6–8 twice.

10 Add 0.1 vol. of sodium acetate to aqueous phase.

11 Add 2 vols of ice-cold, absolute ethanol and mix.

12 Place samples in liquid nitrogen for 1 min. ③

13 Centrifuge for 15 min.

14 Remove the supernatant, being careful not to disturb the pellet, which should be clearly visible. ④

15 Add 1.0 ml of 70% ice-cold ethanol to the tube and mix.

16 Centrifuge for 1 min.

17 Completely remove the supernatant.

18 Leave samples to dry in air for 1 h. ⑤

19 Redissolve samples in 50 μl of 10 mM Tris-HCl pH 7.6, 0.2 mM EDTA. ⑥

④ The pellet should be clearly visible. When small amounts of DNA are being processed, and the pellet is less visible, it should be remembered that the pellet will form on the face of the tube which is furthest from the center of the rotor.

⑤ The samples can be dried under vacuum but such effective drying will require that the pellets be redissolved for much longer times.

⑥ To ensure that the pellet is completely dissolved it is best to leave the samples overnight. Mixing and vortexing can be employed.

Protocol 28. Purification of DNA

Reagents

1 mg/ml BSA
Restriction enzyme buffer (×10)
Restriction enzyme
1 mg/ml RNase A
Termination buffer [20 mM EDTA, 0.4 % (w/v) SDS]

Equipment

Microcentrifuge
Microcentrifuge tubes (1.5 ml)
Pipettes and tips
Water bath

Procedure

1 Combine in a microcentrifuge tube and thoroughly mix the following:
   - substrate DNA   10 µl
   - 10× restriction enzyme buffer   8 µl
   - BSA   8 µl
   - RNase A   4 µl
   - enough enzyme to digest the DNA   1-5 µl
   - distilled water to a final volume of 80 µl   x µl.

2 Incubate at the appropriate temperature for the particular restriction enzyme. Digestion should usually proceed for at least 6 h or overnight.

3 Stop reaction by adding an equal volume of termination buffer.

Notes

Time required: 6–18 hours

① High molecular weight DNA can be difficult to dissolve in small volumes and this procedure calls for restriction digestion in a relatively large volume to allow for redissolving the DNA in a correspondingly large volume.

② In purifying the DNA it is only necessary to carry out one phenol and one chloroform extraction.

③ The protocols given here would provide 40 µg of DNA per sample at this stage. This is two to four times more than would be required for subsequent analysis.

4 Purify DNA by following Steps 2–17 from *Protocol 28.*②

5 Dry DNA under vacuum.

6 Redissolve in 20 μl of appropriate gel loading buffer.③

Protocol 29. Restriction digestion

APPENDIX A: SUPPLIERS

Aldrich Chemical Co Inc., 1001 W St Paul Avenue, PO Box 355, Milwaukee WI 53201, USA.
Tel (414) 263 3850, 800 558 9160. Fax 811 962 959.
The Old Brickyard, New Road, Gillingham, Kent SP8 4XT, UK.
Tel (01747) 822211. Fax (01747) 823779.

Beckman Instruments (UK) Ltd, Oakley Court, Kingsmead Business Park, London Road, High Wycombe, Bucks HP11 1JU, UK.
Tel (01494) 441181. Fax (01494) 463836.

Bio-Rad Laboratories, Bio-Rad House, Maylands Avenue, Hemel Hempstead, Herts HP2 7TD, UK.
Tel (01442) 232522, 0800 181134. Fax (014420) 259118.

CBS Scientific Co. Inc., PO Box 856, Corona del Mar, CA 92014, USA.
Tel (619) 755 4959. Fax (619) 755 4959.

Fuji Medical Systems, Bio-Imaging Systems, PO Box 120035, Stamford 06912, USA.
Tel (203) 393 0300. Fax (203) 327 6485.

Gilson, 3000 W Beltline Highway, PO Box 620027, Middleton WI 53562-0027, USA.
Tel (608) 836 1551. Fax (608) 831 4451.

Invitrogen Corp., 3885 B Sorrento Valley Blvd, San Diego, CA 92121, USA.
Tel 800 655 8288. Fax (619) 597 8201.

Isolab Inc., Drawer 4350, Akron, OH 44321, USA.
Tel (330) 825 4525. Fax (330) 825 8520.

Kodak Clinical Diagnostics Ltd, Mandeville House, 62 The Broadway, Amersham, Bucks, UK.
Tel (01494) 431717. Fax (01494) 725301.

Life Technologies Ltd (Gibco BRL), 3 Fountain Drive, Inchinnan Business Park, Inchinnan PA4 9RF, UK.
Tel (0141) 8146100. Fax (0141) 8146317.

Nunc A/S, Kamstruprej 90, Postbox 280, Kamstrup, DK4000, Denmark.
Tel 46 35 90 65. Fax 46 35 01 05.

Perkin-Elmer Applied Biosystems, 850 Lincoln Center Drive, Foster City, CA 94404-1128, USA.
Tel (415) 570 667. Fax (415) 572 2743.

Perkin-Elmer Applied Biosystems (ABI), Kelvin Close, Birchwood Science Park North, Warrington WA3 7PB, UK.
Tel (01925) 825650. Fax (01925) 282502.

R&D Systems Europe Ltd, 4–10 The Quadrant, Barton Lane, Abingdon, Oxon OX14 3YS, UK.
Tel (01235) 531074. Fax (01235) 533420.

Schleicher and Schuel, PO Box 2012, Keene, NH 03431, USA.
Tel 800 245 4024. Fax (603) 357 3627.

Sigma-Aldrich Co. Ltd, Fancy Road, Poole, Dorset BH12 4QH, UK.
Tel (01202) 733114. Fax (01202) 715460.

Sigma Chemical Co., PO Box 14508, St Louis, MO 63178, USA.
Tel 800 521 8956, (314) 771 5757 (overseas call collect).
Fax 800 325 505, (314) 771 5757 (overseas call collect).

United States Biochemicals (A division of Amersham Life Science), 26111 Miles Road, Cleveland, OH 44122. USA.
Tel (216) 765 5000. Fax (216) 464 5075.

INDEX

THE ESSENTIAL TECHNIQUES SERIES

The *Essential Techniques Series* provides accurate, up-to-date, quality information for the life scientist. These handy pocket-sized manuals are easy to carry, and conveniently spiral-bound making them ideal for lab bench work. *Essential Techniques* books provide value for money by giving all the information required in a single source.

Available in 1996 ...

Antibody Applications
P.J. Delves
0 471 95698 8   September 1995   £14.99/$23.95

Gel Electrophoresis: Nucleic Acids
P. Jones
0 471 96043 8   October 1995   £14.99/$23.95

DNA Isolation and Sequencing
B. Roe, J. Crabtree & A. Kahn
0 471 96324 0   May 1996   £14.99/$23.95

Gel Electrophoresis: Proteins
D.M. Gersten
0 471 96265 1   June 1996   £14.99/$23.95

Gene Transcription: DNA Binding Proteins
Edited by K. Docherty
0 471 97016 6   October 1996   £14.99/$23.95

Gene Transcription: RNA Analysis
Edited by K. Docherty
0 471 96147 7   October 1996   £14.99/$23.95

The above two books are also available as a set:

Gene Transcription
0 471 97097 2   October 1996   £27.50/$44.00

PCR
Edited by J. Burke
0 471 95697 X   October 1996   £14.99/$23.95

THE ESSENTIAL TECHNIQUES SERIES

Also available in 1996 ...

Cell Biology
Edited by D.R. Rickwood & J.R. Harris
0 471 96315 1   due November 1996   approx £14.99/$23.95

Human Chromosome Preparation
D.E. Rooney & B.H. Czepulkowski
0 471 96299 6   due November 1996   approx £14.99/$23.95

Available in 1997 ...

Antibody Production
P.J. Delves
0 471 97010 7   due January 1997   approx £14.99/$23.95

Cell Culture
B. Griffiths & A. Doyle
0 471 97057 3   due April 1997   approx £14.99/$23.95

Vectors: Cloning Applications
P. Gacesa & D. Ramji
0 471 96266 X   due May 1997   approx £14.99/$23.95

Vectors: Expression Systems
P. Gacesa & D. Ramji
0 471 96267 8   due June 1997   approx £14.99/$23.95

Nucleic Acid Hybridization
Edited by J. Ross
0 471 97125 1   due July 1997   approx £14.99/$23.95

ORDER FORM

Please send me:

Qty   Title   Price/copy   Total

_____

_____

_____

All prices are correct at time of going press but subject to change.

Your order will be processed without delay, please allow 21 days for delivery.

We will refund your payment without question if you return any unwanted book to us in a re-saleable condition within 30 days.

All books are available from your bookseller.

Method of payment

☐ Payment £/$ _____ enclosed (Payable to John Wiley & Sons Ltd).

Orders for one book only - please add £2.00/$5.00 to cover postage and handling. Two or more books postage FREE.

☐ Purchase order enclosed ☐ Please send me an invoice
    (£2.00/$5.00 will be added to cover postage and handling).

☐ Please charge my credit card account

☐ American Express ☐ Diners Club ☐ Visa ☐ Mastercard

Card no: _____ Expiry date: _____

Signature: _____

Telephone our Customer Services Dept with your cash or credit card order on (01243) 843206 or dial FREE on 0800 243407 (UK only)

Send my order to:

Name (PLEASE PRINT) _____

Position: _____

Address: _____

_____

_____

_____

Telephone _____

Signature: _____ Date: _____

Return to:

Andrea Sharp, John Wiley & Sons Ltd, Baffins Lane, Chichester, West Sussex, PO19 1UD, UK

Fax: +44 1243 770460

or: Wiley Liss, 605 Third Avenue, New York, NY 10158-0012, USA

Fax: (212) 850 8888

☐ If you do not wish to receive mailings from other companies please tick
    this box or notify the Marketing Services Dept at John Wiley & Sons

WILEY